AF556197

Transportation Engineering

Transportation Engineering

Sateyesh Singh

RANDOM PUBLICATIONS
NEW DELHI (INDIA)

Transportation Engineering

ISBN 978-93-5111-400-0
© Reserved

All Rights Reserved. No Part of this book may be reproduced in any manner without written permission.

Published in 2014 in India by

RANDOM PUBLICATIONS

4376-A/4B, Gali Murari Lal, Ansari Road
New Delhi-110 002
Phone : +91-11-43580356, +91-11-23289044
e-mail: randomexports@gmail.com, sales@randompublications.com, info@randompublications.com

Reprinted 2018

Type Setting by : Keystoneprintads, Delhi-110051
Digitally Printed at : Replika Press Pvt. Ltd.

Preface

Transportation Engineering is the application of science and technology for the safe, efficient and sustainable movement of people and goods. It encompasses research, policy development, planning, design, implementation, operation and management of all modes of travel, be that by road, rail, water or air, and interfaces between these modes and with other land uses. In managing the transport system, it is important to recognize that the overall transport system comprises people, vehicles, infrastructure and communications and other interfaces between these components, and that the transport system is a vital element of the overall land use and economic system. It is also important to understand that transport is a derived demand - it occurs because there is value created when people and goods move from one location to another that exceeds the cost of travel or transport.

Railway, bus, highway, and airport systems are intricate infrastructures that require considerable planning and development. A person who designs these systems is known as a transportation engineer. Transportation engineers may work in a variety of sectors, including consulting firms, private businesses, universities, and government agencies. Depending on his or her specialty, a transportation engineer may be responsible for specific tasks in one concentration only. People who study railroad systems may work only on rail services, while those who concentrate on traffic engineering exclusively may develop new roads and design traffic patterns. A transportation civil engineer who works in the general field may perform all of these tasks, as well as others.

Transportation engineering is the application of technology and scientific principles to the planning, functional design, operation and management of facilities for any mode of transportation in order to provide for the safe, efficient, rapid, comfortable, convenient, economical, and environmentally compatible movement of people and goods (transport). It is a sub-discipline of civil engineering. and of industrial engineering. Transportation engineering is a major component of the civil engineering and mechanical engineering disciplines, according to specialization of academic courses and main competences of the involved territory. The importance of transportation engineering within the civil and industrial engineering profession can be judged by the number of divisions in ASCE (American Society of Civil Engineers) that are directly related to transportation. There are six such divisions (Aerospace; Air Transportation; Highway; Pipeline; Waterway, Port, Coastal and Ocean; and Urban Transportation) representing one-third of the total 18 technical divisions within the ASCE (1987).

The planning aspects of transport engineering relate to urban planning, and involve technical forecasting decisions and political factors. Technical forecasting of passenger travel usually involves an urban transportation planning model, requiring the estimation of trip generation (how many trips for what purpose), trip distribution (destination choice, where is the traveler going), mode choice (what mode is being taken), and route assignment (which streets or routes are being used). More sophisticated forecasting can include other aspects of traveler decisions, including auto ownership, trip chaining (the decision to link individual trips together in a tour) and the choice of residential or business location (known as land use forecasting). Passenger trips are the focus of transport engineering because they often represent the peak of demand on any transportation system.

With pertinent information on such a matter of great relevance and importance, this book would be useful for researchers and students.

I thank all members of my team who have helped in the preparation of the book. My special thanks go to "Random Publications" who have published the book.

– Sateyesh Singh

Contents

1

Transportation

TRANSPORT

Transport or transportation is the movement of people, animals and goods from one location to another. Modes of transportinclude air, rail, road, water, cable, pipeline and space. The field can be divided into infrastructure, vehicles and operations.

Transport is important since it enables trade between people, which in turn establishes civilizations. Transport infrastructure consists of the fixed installations necessary for transport, including roads, railways, airways, waterways,canals and pipelines and terminals such as airports, railway stations, bus stations, warehouses, trucking terminals, refueling depots (including fueling docks and fuel stations) and seaports. Terminals may be used both for interchange of passengers and cargo and for maintenance.

Vehicles traveling on these networks may include automobiles, bicycles, buses, trains, trucks, people, helicopters and aircraft. Operations deal with the way the vehicles are operated, and the procedures set for this purpose including financing, legalities and policies. In the transport industry, operations and ownership of infrastructure can be either public or private, depending on the country and mode. Passenger transport may be public, where operators provide scheduled services, or private. Freight transport has become focused on containerization, although bulk transport is used for large volumes of durable items. Transport plays an important part in economic growth and globalization, but most types cause air pollution and use large amounts of land. While it is heavily subsidized by governments, good planning of transport is essential to make traffic flow and restrain urban sprawl.

MODE

A mode of transport is a solution that makes use of a particular type of vehicle, infrastructure and operation. The transport of a person or of cargo may involve one mode or several of the modes, with the latter case being called intermodal or multimodal transport. Each mode has its own advantages and disadvantages, and will be chosen for a trip on the basis of cost, capability, route and speed.

Human-powered

Human powered transport is the transport of people and/or goods using human muscle-power, in the form of walking, running andswimming. Modern technology has allowed machines to enhance human-power. Human-powered transport remains popular for reasons of cost-saving, leisure, physical exercise and environmentalism. Human-powered transport is sometimes the only type available, especially in underdeveloped or inaccessible regions. It is considered an ideal form of sustainable transportation.

Although humans are able to walk without infrastructure, the transport can be enhanced through the use of roads, especially when using the human power with vehicles, such as bicycles and inline skates. Human-powered vehicles have also been developed for difficult environments, such as snow and water, by watercraft rowing and skiing; even the air can be entered with human-powered aircraft.

Animal-powered

Animal-powered transport is the use of working animals for the movement of people and goods. Humans may ride some of the animals directly, use them as pack animals for carrying goods, or harness them, alone or in teams, to pull sleds or wheeled vehicles.

Air

Fig. An Air France Airbus A318 Lands at London Heathrow Airport

A fixed-wing aircraft, commonly called airplane, is a heavier-than-air craft where movement of the air in relation to the wings is used to generate lift. The term is used to distinguish from rotary-wing aircraft, where the movement of the lift surfaces relative to the air generates lift. A gyroplane is both fixed-wing and rotary-wing. Fixed-wing aircraft range from small trainers and recreational aircraft to large airliners and military cargo aircraft.

Two things necessary for aircraft are air flow over the wings for lift and an area for landing. The majority of aircraft also need anairport with the infrastructure to receive maintenance, restocking, refueling and for the loading and unloading of crew, cargo and passengers. While the vast majority of aircraft land and take off on land, some are capable of take off and landing on ice, snow and calm water. The aircraft is the second fastest method of transport, after the rocket. Commercial jets can reach up to 955 kilometres per hour (593 mph), single-engine aircraft 555 kilometres per hour (345 mph). Aviation is able to quickly transport people and limited amounts of cargo over longer distances, but incur high costs and energy use; for short distances or in inaccessible placeshelicopters can be used. As of April 28, 2009 The Guardian article notes that, "the WHO estimates that up to 500,000 people are on planes at any time."

Rail

Fig. InterCityExpress, a German High-speedpassenger Train

Rail transport is where a train runs along a set of two parallel steel rails, known as a railway or railroad. The rails are anchoredperpendicular to ties (or sleepers) of timber, concrete or steel, to maintain a consistent distance apart, or gauge. The rails and perpendicular beams are placed on a foundation made of concrete, or compressed earth and gravel in a bed of ballast. Alternative methods include monorail and maglev.

A train consists of one or more connected vehicles that run on the rails. Propulsion is commonly provided by a locomotive, that hauls a series of unpowered cars, that can carry passengers or freight. The locomotive can be powered by steam, diesel or byelectricity supplied by trackside systems. Alternatively, some or all the cars can be powered, known as a multiple unit. Also, a train can be powered by horses, cables, gravity, pneumatics and gas turbines. Railed vehicles move with much less friction than rubber tires on paved roads, making trains more energy efficient, though not as efficient as ships.

Intercity trains are long-haul services connecting cities; modern high-speed rail is capable of speeds up to 350 km/h (220 mph), but this requires specially built track. Regional and commuter trains feed cities from suburbs and surrounding areas, while intra-urban transport is performed by high-capacity tramways and rapid transits, often making up the backbone of a city's public transport. Freight trains traditionally used box cars, requiring manual loading and unloading of the cargo. Since the 1960s, container trains have become the dominant solution for general freight, while large quantities of bulk are transported by dedicated trains.

Road

Fig. Interstate 80 Near Berkeley, California, United States.

A road is an identifiable route, way or path between two or more places. Roads are typically smoothed, paved, or otherwise prepared to allow easy travel; though they need not be, and historically many roads were simply recognizable routes without any formal construction or maintenance. In urban areas, roads may pass through a city or village and be named as streets, serving a dual function as urban space easement and route. The most common road vehicle is the automobile; a wheeled passenger vehicle that carries its own motor. Other users of roads include buses, trucks, motorcycles, bicycles and pedestrians. As of 2002, there were 590 million automobiles worldwide. Automobiles offer high flexibility and with low capacity, but are deemed with high energy and area use, and the main source ofnoise and air pollution in cities; buses allow for more efficient travel at the cost of reduced flexibility. Road transport by truck is often the initial and final stage of freight transport.

Water

Fig. Built by the Dutch to Transport Spices, now Used by the Local Fisherman to get to the Sea, Negombo Dutch Canal, Sri Lanka

Fig. Automobile Ferry in Croatia

Water transport is movement by means of a watercraft—such as a barge, boat, ship orsailboat—over a body of water, such as a sea, ocean, lake, canal or river. The need for buoyancy is common to watercraft, making the hull a dominant aspect of its construction, maintenance and appearance. In the 19th century the first steam ships were developed, using a steam engine to drive a paddle wheel or propeller to move the ship. The steam was produced in a boilerusing wood or coal and fed through a steam external combustion engine. Now most ships have an internal combustion engine using a slightly refined type of petroleumcalled bunker fuel. Some ships, such as submarines, use nuclear power to produce the steam. Recreational or educational craft still use wind power, while some smaller craft use internal combustion engines to drive one or more propellers, or in the case of jet boats, an inboard water jet. In shallow draft areas, hovercraft are propelled by large pusher-prop fans.

Although slow, modern sea transport is a highly efficient method of transporting large quantities of goods. Commercial vessels, nearly 35,000 in number, carried 7.4 billion tons of cargo in 2007. Transport by water is significantly less costly than air transport for transcontinental shipping; short sea shipping and ferries remain viable in coastal areas.

Other Modes

Fig. Trans-Alaska Pipeline for Crude Oil

Pipeline transport sends goods through a pipe, most commonly liquid and gases are sent, but pneumatic tubes can also send solid capsules using compressed air. For liquids/gases, any chemically stable liquid or gas can be sent through a pipeline. Short-distance systems exist for sewage, slurry, water and beer, while long-distance networks are used for petroleum and natural gas. Cable transport is a broad mode where vehicles are pulled by cables instead of an internal power source. It is most commonly used at steep gradient. Typical solutions include aerial tramway, elevators, escalator and ski lifts; some of these are also categorized asconveyor transport.

Spaceflight is transport out of Earth's atmosphere into outer space by means of a spacecraft. While large amounts of research have gone into technology, it is rarely used except to put satellites into orbit, and conduct scientific experiments. However, man has landed on the moon, and probes have been sent to all the planets of the Solar System. Suborbital spaceflight is the fastest of the existing and planned transport systems from a place on Earth to a distant other place on Earth. Faster transport could be achieved through part of a Low Earth orbit, or following that trajectory even faster using the propulsion of the rocket to steer it.

ELEMENTS

Infrastructure

Fig. Bridges, such as Golden Gate Bridge, Allow Roads and Railways to Cross Bodies of Water

Infrastructure is the fixed installations that allow a vehicle to operate. It consists of a way, a terminal and facilities for parking and maintenance. For rail, pipeline, road and cable transport, the entire way the vehicle travels must be built up. Air and water craft are able to avoid this, since the airway and seaway do not need to be built up. However, they require fixed infrastructure at terminals. Terminals such as airports, ports and stations, are locations where passengers and freight can be transferred from one vehicle or mode to another. For passenger transport, terminals are integrating different modes to allow riders to interchange to take advantage of each mode's advantages. For instance, airport rail links connect airports to the city centers and suburbs.

The terminals for automobiles are parking lots, while buses and coaches can operates from simple stops. For freight, terminals act as transshipment points, though some cargo is transported directly from the point of production to the point of use.

The financing of infrastructure can either be public or private. Transport is often a natural monopoly and a necessity for the public; roads, and in some countries railways and airports are funded through taxation. New infrastructure projects can involve large spendings, and are often financed through debt. Many infrastructure owners therefore impose usage fees, such as landing fees at airports, or toll plazas on roads. Independent of this, authorities may impose taxes on the purchase or use of vehicles.

Vehicles

Fig. A Fiat Uno in 2008

A vehicle is any non-living device that is used to move people and goods. Unlike the infrastructure, the vehicle moves along with the cargo and riders. Unless being pulled by a cable or muscle-power, the vehicle must provide its own propulsion; this is most commonly done through a steam engine, combustion engine, electric motor, a jet engine or a rocket, though other means of propulsion also exist.

Vehicles also need a system of converting the energy into movement; this is most commonly done throughwheels, propellers and pressure. Vehicles are most commonly staffed by a driver. However, some systems, such as people movers and some rapid transits, are fullyautomated. For passenger transport, the vehicle must have a compartment for the passengers. Simple vehicles, such as automobiles, bicycles or simple aircraft, may have one of the passengers as a driver.

Operation

Private transport is only subject to the owner of the vehicle, who operates the vehicle themselves. For public transport and freight transport, operations are done through private enterprise or by governments. The infrastructure and vehicles may be owned and operated by the same company, or they may be operated by different entities.

Fig. Incheon International Airport, South Korea

Traditionally, many countries have had a national airline and national railway. Since the 1980s, many of these have been privatized. International shipping remains a highly competitive industry with little regulation, but ports can be public owned.

FUNCTION

Relocation of travellers and cargo are the most common uses of transport. However, other uses exist, such as the strategic and tactical relocation of armed forces during warfare, or the civilian mobility construction or emergency equipment.

Passenger

Fig. A Local Transit Bus Operated by ACTION in Canberra, Australia

Passenger transport, or travel, is divided into public and private transport. Public transport is scheduled services on fixed routes, while private is vehicles that provide ad hoc services at the riders desire. The latter offers better flexibility, but has lower capacity, and a higher environmental impact. Travel may be as part of daily commuting, for business, leisure or migration. Short-haul transport is dominated by the automobile and mass transit. The latter consists of buses in rural and small cities, supplemented with commuter rail, trams and rapid transit in larger cities. Long-haul transport involves the use of the automobile, trains, coaches and aircraft, the last of which have become predominantly used for the longest, including intercontinental, travel.Intermodal passenger transport is where a journey is performed through the use of several modes of transport; since all human transport normally starts and ends with walking, all passenger transport can be considered intermodal. Public transport may also involve the intermediate change of vehicle, within or across modes, at a transport hub, such as a bus or railway station.

Taxis and Buses can be found on both ends of Public Transport spectrum, whereas Buses remain the cheaper mode of transport but are not necessarily flexible, and Taxis being very flexible but more expensive. In the middle is Demand responsive transportoffering flexibility whilst remaining affordable. International travel may be restricted for some individuals due to legislation and visa requirements.

Freight

Freight transport, or shipping, is a key in the value chain in manufacturing. With increased specialization and globalization, production is being located further away from consumption, rapidly increasing the demand for transport. While all modes of transport are used for cargo transport, there is high differentiation between the nature of the cargo transport, in which mode is chosen. Logistics refers to the entire process of transferring products from producer to consumer, including storage, transport, transshipment, warehousing, material-handling and packaging, with associated exchange of information. Incoterm deals with the handling of payment and responsibility ofrisk during transport.

Fig. Freight Train with Shipping Containers in the United Kingdom.

Containerization, with the standardization of ISO containers on all vehicles and at all ports, has revolutionized international anddomestic trade, offering huge reduction in transshipment costs. Traditionally, all cargo had to be manually loaded and unloaded into the haul of any ship or car; containerization allows for automated handling and transfer between modes, and the standardized sizes allow for gains in economy of scale in vehicle operation. This has been one of the key driving factors in international tradeand globalization since the 1950s.

Bulk transport is common with cargo that can be handled roughly without deterioration; typical examples are ore, coal, cereals andpetroleum. Because of the uniformity of the product, mechanical handling can allow enormous quantities to be handled quickly and efficiently. The low value of the cargo combined with high volume also means that economies of scale become essential in transport, and gigantic ships and whole trains are commonly used to transport bulk. Liquid products with sufficient volume may also be transported by pipeline. Air freight has become more common for products of high value; while less than one per cent of world transport by volume is by airline, it amounts to forty per cent of the value. Time has become especially important in regards to principles such aspostponement and just-in-time within the value chain, resulting in a high willingness to pay for quick delivery of key components or items of high value-to-weight ratio. In addition to mail, common items sent by air include electronics and fashion clothing.

HISTORY

Humans' first means of transport were walking and swimming. The domestication of animals introduces a new way to lay the burden of transport on more powerful creatures, allowing heavier loads to be hauled, or humans to ride the animals for higher speed and duration. Inventions such as the wheel and sled helped make animal transport more efficient through the introduction of vehicles. Also water transport, including rowed and sailed vessels, dates back to time immemorial, and was the only efficient way to transport large quantities or over large distances prior to the Industrial Revolution.

The first forms of road transport were horses, oxen or even humans carrying goods over dirt tracks that often followed game trails. Paved roads were built by many early civilizations, including Mesopotamia and the Indus Valley Civilization. The Persian andRoman empires built stone-paved roads to allow armies to travel quickly. Deep roadbeds of crushed stone underneath ensured that the roads kept dry. The medieval Caliphate later built tar-paved roads. The first watercraft were canoes cut out from tree trunks. Early water transport was accomplished with ships that were either rowed or used the wind for propulsion, or a combination of the two. The importance of water has led to most cities, that grew up as sites for trading, being located on rivers or at sea, often at the intersection of two bodies of water. Until the Industrial Revolution, transport remained slow and costly, and production and consumption were located as close to each other as feasible. The Industrial

Revolution in the 19th century saw a number of inventions fundamentally change transport. With telegraphy, communication became instant and independent of transport. The invention of the steam engine, closely followed by its application in rail transport, made land transport independent of human or animal muscles. Both speed and capacity increased rapidly, allowing specialization through manufacturing being located independent of natural resources. The 19th century also saw the development of the steam ship, that sped up global transport.

With the development of the combustion engine and the automobile at the turn into the 20th century, road transport became more viable, allowing the introduction of mechanical private transport. The first highways were constructed during the 19th century withmacadam. Later, tarmac and concrete became the dominant paving material. In 1903, the first controllable airplane was invented, and after World War I, it became a fast way to transport people and express goods over long distances.

After World War II, the automobile and airlines took higher shares of transport, reducing rail and water to freight and short-haul passenger. Spaceflight was launched in the 1950s, with rapid growth until the 1970s, when interest dwindled. In the 1950s, the introduction of containerization gave massive efficiency gains in freight transport, permitting globalization. International air travel became much more accessible in the 1960s, with the commercialization of thejet engine. Along with the growth in automobiles and motorways, this introduced a decline for rail and water transport. After the introduction of the Shinkansen in 1964, high-speed rail in Asia and Europe started taking passengers on long-haul routes from airlines.

Early in U.S. history, most aqueducts, bridges, canals, railroads, roads, and tunnels were owned by private joint-stock corporations. Most such transportation infrastructure came under government control in the late 19th and early 20th centuries, culminating in the nationalization of inter-city passenger rail service with the creation of Amtrak. Recently, however, a movement to privatize roads and other infrastructure has gained some ground and adherents.

IMPAC

Economic

Transport is a key necessity for specialization—allowing production and consumption of products to occur at different locations. Transport has throughout history been a spur to expansion; better transport allows more trade and a greater spread of people.Economic growth has always been dependent on increasing the capacity and rationality of transport. But the infrastructure and operation of transport has a great impact on the land and is the largest drainer of energy, making transport sustainability a major issue.

Fig. Transport is a Key Component of Growth and Globalization, such as in Seattle, Washington, United States

Modern society dictates a physical distinction between home and work, forcing people to transport themselves to places of work or study, as well as to temporarily relocate for other daily activities. Passenger transport is also the essence of tourism, a major part of recreational transport. Commerce requires the transport of people to conduct business, either to allow face-to-face communication for important decisions or to move specialists from their regular place of work to sites where they are needed.

Planning

Transport planning allows for high utilization and less impact regarding new infrastructure. Using models of transport forecasting, planners are able to predict future transport patterns. On the operative level, logistics allows owners of cargo to plan transport as part of the supply chain. Transport as a field is studied through transport economics, the backbone for the creation of regulation policy by authorities. Transport engineering, a sub-discipline of civil engineering, and must take into account trip generation, trip distribution, mode choice and route assignment, while the operative level is handled through traffic engineering.

Fig. The Engineering of this Roundabout in Bristol, United Kingdom, Attempts to make Traffic Flow Free-moving

Because of the negative impacts made, transport often becomes the subject of controversy related to choice of mode, as well as increased capacity. Automotive transport can be seen as a tragedy of the commons, where the flexibility and comfort for the individual deteriorate the natural and urban environment for all. Density of development depends on mode of transport, with public transport allowing for better spacial utilization. Good land use keeps common activities close to peoples homes and places higher-density development closer to transport lines and hubs; minimize the need for transport. There are economies of agglomeration. Beyond transportation some land uses are more efficient when clustered. Transportation facilities consume land, and in cities, pavement (devoted to streets and parking) can easily exceed 20 per cent of the total land use. An efficient transport system can reduce land waste.

Too much infrastructure and too much smoothing for maximum vehicle throughput means that in many cities there is too much traffic and many—if not all—of the negative impacts that come with it. It is only in recent years that traditional practices have started to be questioned in many places, and as a result of new types of analysis which bring in a much broader range of skills than those traditionally relied on—spanning such areas as environmental impact analysis, public health, sociologists as well as economists who increasingly are questioning the viability of the old mobility solutions. European cities are leading this transition.

Environment

Fig. Traffic Congestion Persists in São Paulo, Brazil Despite the No-drive Days Based on License Numbers.

Transport is a major use of energy and burns most of the world's petroleum. This creates air pollution, including nitrous oxides andparticulates, and is a significant contributor to global warming through emission of carbon dioxide, for which transport is the fastest-growing emission sector. By subsector, road transport is the largest contributor to global warming. Environmental regulations in developed countries have reduced individual vehicles' emissions; however, this has been offset by increases in the numbers of vehicles and in the use of each vehicle. Some pathways to reduce the carbon emissions of road vehicles considerably have been studied. Energy use and emissions vary largely between modes, causing environmentalists to call for a transition from air and road to rail and human-powered transport, as well as increased transport electrification and energy efficiency.

Other environmental impacts of transport systems include traffic congestion and automobile-oriented urban sprawl, which can consume natural habitat and agricultural lands. By reducing transportation emissions globally, it is predicted that there will be significant positive effects on Earth's air quality, acid rain, smog and climate change.

PURPOSE AND OVERVIEW OF TRANSPORTATION

The Transportation Element of the Plan provides a framework for the rational and orderly development of the area's street system, including Expressways, Arterials, Collectors, and Local Streets. This plan element includes a review of previous transportation plans and studies, overview of existing transportation facilities and services, analysis of travel characteristics, and recommendations for development of the thoroughfare system plan for the Harlingen area. Harlingen's thoroughfare system plan is shown in the Long Range Thoroughfare Plan, which is the City's general plan for extension and widening of the roads, streets, and public highways within the municipality's Corporate Limits and its Extraterritorial Jurisdiction.

TRANSPORTATION PLANNING AREA

This Transportation Element addresses the same geographic area as other elements of the Comprehensive Plan, which is the area encompassed by the City of Harlingen and its Extraterritorial Jurisdiction (ETJ). The planning area boundary is illustrated in Figure, shown previously in Chapter 1 - Introduction.

Other Related Transportation Plans and Studies

There are a number of related plans and studies that have addressed transportation improvement needs for the City of Harlingen and the Harlingen-San Benito Metropolitan Area, which were reviewed and considered in the development and update of this Comprehensive Plan element. These relevant plans and studies are highlighted following.

METROPOLITAN TRANSPORTATION PLAN

Harlingen-San Benito Metropolitan Transportation Plan (December 1999) - The Harlingen-San Benito Metropolitan Planning Organization (MPO), in cooperation with the Texas Department of Transportation, the Federal Transit Administration, and Federal Highway Administration, is responsible for regional transportation planning for the Harlingen-San Benito urbanized area. Prepared by MPO staff, the Metropolitan Transportation Plan is the area's 25-year long-range transportation improvement plan, which identifies the planned transportation improvements that are eligible to use Federal funding. Approximately 150 specific roadway improvement projects are included in the plan. It is financially constrained, meaning that the estimated implementation cost of the planned improvements is consistent with the amounts of Federal and State funding expected to be available during the life span of the plan.

Railroad Crossing Study

Harlingen-San Benito Railroad Crossing Study (April 1998) This Study was prepared for the MPO by Traffic Engineers, Inc., the study developed recommendations to improve the safety of the at-grade railroad crossings and alleviate vehicular congestion and delays. The recommendations designed to improve the safety of the at-grade railroad crossings include signing and pavement markings; re-planking; signal preemption; installation of signals and/or gates; and consolidation of crossings.

Assessment of Public Transportation Needs and Transit Plan

Assessment of Public Transportation Needs and Transit Plan (June 1996) This study was prepared by LKC Consulting Services in association with Michael J. Blum Company and Traffic Engineers, Inc. It was done to determine if there was a need for public transit services in the Harlingen-San Benito metropolitan area. It identified target populations in need of additional transportation services, and identified transit alternatives feasible to meet the needs of the area. Conclusions were that the population density was not sufficient to support a fixed route transit system but that indicators of transit needed to be monitored.

Cameron County Thoroughfare Plan

Cameron County Thoroughfare Plan (1994) - This thoroughfare plan for Cameron County was prepared by the County Engineer's Office. The plan identifies existing major roads and highways in the county and shows general alignments for new thoroughfares. Improvements were included in a Road Improvement Bond Program being implemented by Cameron County.

Valle Vista Mall and Haverford Area Traffic Study

Traffic Study - Valle Vista Mall and Haverford Area (January 1994) - Commercial and residential growth in the Valle Vista Mall and Haverford area have generated high traffic demand on arterial roadways. This study prepared for the MPO by Traffic Engineers, Inc., recommended improvements to Dixieland Rd., Lincoln Ave., and the US 83. Interchange at M St., and the

US 77/83 Ramps, as well providing efficient and safe access to Harlingen South High School and the Haverford residential area.

Mass Transit Study

Harlingen/South Padre Island Mass Transit System Study (June 1993) - Prepared for the Lower Rio Grande Valley Development Council by Lichliter/ Jameson & Associates, this study evaluated public transportation needs of the Harlingen-South Padre Island corridor with the purpose of identifying the feasibility of a mass transit system. This study recommended US 77/83 be widened to six lanes; FM 510 resurfaced; FM 3462 extended to link Loop 590 to SH 345; and, further consideration given to the development of a regional mass transit system.

Airport Noise Study

Valley International Airport FAR Part 150 Noise Exposure and Land Use Compatibility Program (September 1992) - This is a five-year program developed locally and approved by the Federal Aviation Administration, which assesses the airport noise environment, forecasts aviation operations, identifies land uses within the airport environs, and recommends ways to mitigate land use compatibility conflicts.

Valley Mobility Plan

Rio Grande Valley Mobility Plan (July 1992) - The Rio Grande Valley Chamber of Commerce sponsored the preparation of this regional mobility plan for Cameron, Hidalgo, Starr, and Willacy Counties. Shiner Mosley & Associates developed the plan for the Chamber of Commerce. Transportation improvement needs identified for the Harlingen area included the Harlingen Crosstown Expressway, Inner Loop (Loop 499), SH 345/FM 106, US 77/83, Loop 590, Business 77, SH 107, FM 1479, and West Arterial project groups.

Airport Master Plan

Valley International Airport Master Plan (December 1991) - Prepared for the City of Harlingen Airport Board by Barnard Dunkelberg & Company, the Airport Master Plan is the City's plan for future development of Valley International Airport. The 20-year plan identified 45 proposed improvement projects estimated to cost approximately $46.5 million in 1991 dollars. An update is currently underway by the same firm who completed the original document. Estimated completion date is December 2000.

West Arterial Study

Arterial Alignment Study - Dixieland Rd. to Bass Blvd. (September 1991) - Prepared for the MPO by Wilbur Smith Associates, this study evaluated alternative corridor alignments for an east-west arterial extending between

Dixieland Road and Bass Boulevard (FM 800), southwest of Harlingen. The study recommended the construction of a continuous West Arterial, generally following the alignments of Grand Road, Garrett Road and Regina Lane. The recommended arterial should be constructed as a two-lane roadway with a right-of-way width of 120 feet, which would accommodate an ultimate four-lane (or six-lane) urban roadway. This study also recommended future development of another arterial roadway south of and generally parallel to the recommended West Arterial.

Population and Socio-economic Data

Transportation Plan Update for Harlingen-San Benito Urbanized Area (September 1990) - This MPO study prepared by Wilbur Smith Associates updated the demographic and socio-economic data for the Harlingen-San Benito Urban Transportation Study area, including 1989 base year data and forecast years 2000 and 2010 data. These data consisted of population, dwelling units, median household income, basic employment, retail employment, service employment, and special generators for traffic analysis zones (TAZ's) within the metropolitan area.

Expressway Ramp Analysis

US 77/83 Ramp Analysis (April 1990) - The expressway ramp analysis of US 77/83 identified existing and future traffic service deficiencies and needed improvements. The MPO study was performed by Wilbur Smith Associates and recommended the following improvements: completion of the US 77/83 at Ed Carey Dr. interchange traffic signalization; widening FM 801 west of the expressway; implementing a standard diamond interchange traffic signal phasing on the Sam Houston Blvd. interchange; upgrading ramps and widening of the Ed Carey Dr., Helen Moore and Williams Rd. interchanges; extend FM 510 with grade separation at the expressway; and the conversion of frontage roads from two-way to one-way between McCullough to FM 732.

Traffic Engineering Studies

Traffic Engineering Studies at Various Intersections in the City of Harlingen, Texas, Work Order #87-07-21 (October 1988) - Traffic engineering studies were conducted by Wilbur Smith Associates at the intersection of westbound Business US 83 (Harrison Ave.) and Dixieland Rd., and the adjacent intersection of eastbound Business US 83 and Dixieland.

These studies recommended the installation of a two-phase traffic signal (providing coordination with signal at Business US 83 Eastbound) and overhead guide signage for the intersection of Business US 83 Westbound and Dixieland Road. It also recommended supplemental signal faces for the eastbound approach on the intersection of Business US 83 Eastbound and Dixieland Road.

Downtown Traffic and Parking Study

Downtown Parking and Traffic Flow Study (April 1988) - This study by De Shazo, Starek & Tang evaluated the existing traffic and parking conditions in central Harlingen and provided recommendations for a downtown parking and traffic flow plan. In addressing parking conditions this study recommended: standardized parking spaces and meters, increased parking fees and fines, development of new parking lots, adding of curb parking along Monroe St., and implementing of parking requirement policies. Although traffic conditions at the time were considered adequate, it recommended widening First St. to four lanes and closing A Street at its intersection with Van Buren Ave. This study also recommended the implementation of transportation related amenities such as "corner knuckles" along Jackson Ave., "pocket parks" at the intersections of A Street/Van Buren Ave. and B Street/ Commerce St., and addition of street trees and plantings to enhance the downtown pedestrian environment.

Arroyo Crossings Study

Arroyo Colorado Crossings Study (July 1987) - The Arroyo Colorado Crossings Study re-examined the need for new roadway crossings based on existing (1987) and future (2005) traffic conditions and recommended crossing alignments to preserve the right-of-way for future implementation. Prepared for the MPO by Henningson, Durham & Richardson, this study recommended three Arroyo crossing location alternatives. The first alternative extends Morris Rd. to the intersection of Mayfield Rd. and Dixieland Rd. The second alternative connects the intersection of FM 1595 and FM 106 with Helen Moore Rd. The third alternative parallels the FM 1846/FM 106 crossing beginning at FM 106 and aligning with Cemetery Rd. at FM 1846.

Population and Land Use Data

Year 2005 Transportation Plan Update (July 1987) - This plan update developed Year 2005 population and land use tables for the Harlingen-San Benito area, which were needed for updating 20-year traffic projections and evaluating additional Arroyo Colorado crossings. Henningson, Durham & Richardson performed the study for the MPO.

Long-Range Transportation Plan

Harlingen-San Benito Urban Transportation Study Year 2000 Street and Highway Plan, 10-Year Improvement Program (July 1978) - The primary purpose of this study was to develop an update of the area's long-range transportation plan. The study was sponsored by the MPO and performed by Henningson, Durham & Richardson. The proposed arterial projects consisted of the following: extension of Tyler St., Lela St., Thirteenth St., Taft Ave., FM 510 (Loop 448 to US 77/83), and FM 509 to the International Airport;

widening of Valley Fair Blvd., Seventh St., La Palma Blvd. (Combes Rd. to Oscar Williams Rd.), New Combes Rd. (Matz Ave. to Loop 499), Ed Carey Blvd. Loop 374 (Palm Dr. to Stuart Place Rd., and First St. north of Harlingen CBD; improve continuity for Wilson Rd. and Markowsky Ave., Jefferson Ave., Loop 499 at Primera Rd., Turner St. (Sam Houston Blvd.-Ratliff Rd.-Oscar Williams Rd.), La Palma Blvd. to intersect Shafer Rd., FM 509 (north of FM 800 to FM 1846 extension), Helen Moore Rd. (Pennsylvania Ave. to Loop 448), and the Pennsylvania-Morris-Dixieland corridor; and additional capacity improvements and/or operation changes to Loop 448 bridge, Loop 499, Dixieland Rd., Robertson/Zaragosa couplet from Combes Rd. to Reagan St., and US 77/83. In addition, this plan recommended the construction of at least one new crossing of the Arroyo Colorado.

RELATIONSHIP OF TRANSPORTATION ELEMENT AND METROPOLITAN TRANSPORTATION PLAN

Updates of the Transportation Element coincide with development of the Metropolitan Transportation Plan updates for the Harlingen-San Benito Area by the Metropolitan Planning Organization (MPO). The MPO is a cooperative transportation planning organization made up of representatives of State and local governmental entities. The Metropolitan Transportation Plan addresses roadway and transportation improvement needs for the metropolitan area based upon projected travel demands for a 25-year planning period (currently 2000 to 2025). The MPO transportation plan focuses on the allocation of transportation improvement funds available from Federal, State and local sources to meet identified needs over the 25-year period. The Transportation Element of the Comprehensive Plan is the City of Harlingen's thoroughfare system development plan. While the MPO's Metropolitan Transportation Plan and the City's Transportation Element of the Comprehensive Plan are mutually consistent and share many common goals, objectives, and planned improvements, it is important to recognize the following differences between the two plans:

- The two plans were prepared for different but related purposes, since the MPO plan identifies specific improvement needs and planned improvements on roadways which are classified by TxDOT as a collector and above, while Harlingen's Comprehensive Plan identifies general alignments and right-of-way requirements for future thoroughfare system development;
- The two planning areas are different since the MPO plan addresses the urbanized area of Harlingen and San Benito, while the Comprehensive Plan addresses the area encompassed within Harlingen-s City Limits and 5-mile ETJ;
- The planning periods are different since the MPO plan has a time span of 2000-2025, while the Comprehensive Plan extends to the

year 2020 and beyond, including very long-range needs for ensuring preservation of the rights-of-way required for future thoroughfare improvements; and,

- The plans were developed under different enabling and implementing legislation, since the MPO plan responds to TEA 21, the Transportation Equity Act for the 21st Century, which builds upon the U.S. Intermodal Surface Transportation Efficiency Act (ISTEA) of 1991 (Title 23, U.S.C.), and the Comprehensive Plan is governed by the Texas Local Government Code and Harlingen's City Charter.

Constraints To Thoroughfare Development

The Harlingen area has several natural and man-made barriers that were considered in developing the Transportation Element of the Comprehensive Plan. Major constraints influencing development of roads, streets and highways include the existing expressways, Arroyo Colorado, Resaca de Los Fresnos, North Floodway, drainage channels, irrigation canals, and the railroads traversing the area.

The existing railroads and train traffic are a major constraint for east-west movement of automobile traffic in the Harlingen area. At-grade intersections of the railroad lines with area roadways are a cause of traffic delays and traffic safety concerns. Rerouting the railroad lines to new alignment outside the densely developed area is a recommended solution. Other alternatives include constructing grade separated over or underpasses at major railroad-roadway intersections, and traffic safety improvements at existing crossings.

Other constraints to thoroughfare system development include steep slopes along banks of the Arroyo, lakes and reservoirs, floodplain areas, public parks, agricultural lands, wetlands and other critical habitat areas, the Rio Grande Valley International Airport, and Harlingen Industrial Parks. Existing residential neighborhoods and other developed areas also present constraints when a thoroughfare improvement might impact these areas. It is not intended to imply that these existing constraints prohibit the development of a desirable transportation system. However, their influence may affect the feasibility, location, and construction cost of transportation improvements and they should be considered in planning and design of future facilities.

Authority for Planning and Regulating Thoroughfares

Under the provisions of Article XI, Section 5 of the Texas Constitution and Title 7, Chapter 212 of the Texas Local Government Code, the City of Harlingen may require that development plans and subdivision plats must conform to "...the general plan of the municipality and its current and future streets..." and, "...the general plan for extension of the municipality and its roads, streets, and public highways within the municipality and its extraterritorial jurisdiction..." The City of Harlingen's Municipal Charter also contains provisions relating to regulation of plats and thoroughfare development.

Requirements for right-of-way dedication and construction of street improvements apply to all subdivision of land within the City's incorporated area and its extraterritorial jurisdiction (ETJ). In accordance with the Texas Local Government Code, the City has adopted rules governing plats and subdivision of land within the municipality's jurisdiction, and, by ordinance; those rules have also been extended to the city's ETJ.

Functional Classification of Thoroughfares

Thoroughfare Classification Criteria Thoroughfares are grouped into functional classes according to the character of service they are intended to provide. Thoroughfares are classified according to their functional role in terms of movement and access.

The functional classification of thoroughfares is shown by Harlingen's Long Range Thoroughfare Plan and includes the following six functional classes:

- Freeway/Expressways;
- Principal Arterials;
- Minor Arterials;
- Major Collectors;
- Minor Collectors; and,
- Local Streets.

Thoroughfare Classification Criteria

Criteria used in determining the functional classifications of thoroughfares. Classification is based on each roadway's functional role in the network, and the existing and future travel patterns and areas served. The functional classification of a thoroughfare normally does not change as traffic increases and improvements are made. Functional classification is not necessarily related to the number of lanes, although higher classes tend to be multi-lane roadways. However, a two-lane roadway may function as a Principal Arterial in developing or rural areas.

Freeways and Expressways

Freeways/Expressways are devoted entirely to traffic movement with little or no direct land service function. Freeways are multi-lane divided highways with full access control, meaning that all intersections are grade separated and the main lanes provide no direct access to adjoining properties. This class would include Interstate Highways and tollways. There are currently no Interstate Highways or toll roads in Harlingen. Expressways are multi-lane divided roadways with a high degree of partial access control, meaning few, if any, intersections at grade. Full or partial control of access distinguishes Freeways/Expressways from other classes of roadways. Freeways and Expressways serve large volumes of high speed traffic and are primarily intended to serve long trips, including both vehicles entering and leaving the urban area, as well as major circulation movements within the urban area. In the Harlingen area, the roadway facilities that function as freeways and expressways are US 77, US 83 and US 77/83.

Although the Freeway/Expressway classification includes the main lanes of designated facilities, continuous frontage roads (also termed service roads or feeders) are more appropriately classified as Minor Arterials or Major Collectors. It is desirable to have another parallel Principal or Minor Arterial offset one-half to one mile along either side of a Freeway or Expressway, to provide circulation for traffic movement along one-way frontage roads.

Principal Arterials

Principal Arterials are streets and highways that provide a high degree of mobility, serve relatively high traffic volumes, have high operational speeds, and serve a significant portion of through travel or long distance trips. Freeways and Principal Arterials together typically accommodate about 30 to 40 per cent of a region's travel on 5 to 10 per cent of the total roadway network. Principal Arterials serve as primary routes through the Harlingen area and between major destinations within the area. They are continuous over long distances and serve trips entering and leaving the urban area, as well as trips within it. These facilities generally serve high-volume travel corridors that connect major generators of traffic, such as the central business district, other large employment centers, suburban commercial centers, industrial centers, major residential communities, and other major activity centers within the urban area. Some examples of Principal Arterials in the Harlingen area include Business 83, Business 77, Loop 499, Loop 509, and Tyler/Harrison Avenues.

Lower volume roadways that are continuous over long distances may also function as Principal Arterials, particularly in fringe and rural areas. The cross section of Principal Arterials may vary from multi-lane roadways with three, four, five or more lanes, down to two-lane roadways in developing fringe and rural areas where traffic volumes have not increased to the point that more travel lanes are needed. Functional classification is not dependent on the existing number of lanes, since the functional role served by a roadway typically remains constant over time, while the roadway's cross section is improved to accommodate increasing traffic volumes.

Principal Arterials form an interconnecting network for citywide and regional movement of traffic, including connections to Freeways/Expressways serving the region, and to Minor Arterials and Collectors. A two-mile spacing is generally desirable between Principal Arterials, with a one-mile spacing between a Principal Arterial and a Minor Arterial or Freeway. Since traffic movement, not land access, is the primary function of Principal Arterials, access management is essential. Driveways connecting directly onto a Principal Arterial should be minimized to avoid traffic congestion and delays caused by turning movements for vehicles entering and exiting driveways. Off-peak travel speeds on Principal Arterials are typically 40-55 MPH, and peak period speeds are about 30-35 MPH. Intersections with other public streets and private access should be designed to limit speed differentials between turning vehicles and other traffic to no more than 10-15 MPH. Signalized intersection spacing should be long enough to allow a variety of signal cycle lengths and timing plans, which can be adjusted to meet changes in traffic volumes and maintain traffic progression (desirably one-third to one-half mile spacing).

Minor Arterials

Minor Arterials are similar in function to Principal Arterials, except that they provide a higher degree of local access than Principal Arterials does. Minor Arterials include all remaining arterial streets and highways in the urban area and serve less concentrated traffic-generating areas such as neighborhood shopping centers and employment centers. Although Minor Arterials are very similar in function to Principal Arterials, this class typically distributes medium traffic volumes for shorter distance trips than Principal Arterials.

In general, the projected future traffic volumes on Minor Arterials will be lower than the volumes carried by Principal Arterials. Minor Arterials are generally continuous over shorter distances than Principal Arterials. Travel speeds along Minor Arterials are typically 30 to 45 miles per hour in off-peak periods, and 20 to 35 miles per hour in peak periods. Minor Arterial streets serve as boundaries to neighborhoods and collect traffic from Collectors and Local Streets. Although a Minor Arterial typically provides more local access than a Principal Arterial does, the primary function is still traffic movement.

Minor Arterials are distinguished from Principal Arterials by shorter travel distances, slightly lower traffic volumes, and somewhat lower speeds. In areas where closely spaced Arterials are nearly equivalent in functional role, differentiation between Principal and Minor Arterials often involves selection among roadways with similar traffic volumes and physical characteristics. Choices should be made as to whether both roadways should be developed to serve equal functions, or one should be developed to serve a higher function than the other. In such instances, the delineation of classes includes consideration of the most efficient plan for future improvements. Principal and Minor Arterials are generally spaced at one mile intervals in an alternating grid pattern. The integrated system formed by Principal Arterials and Minor Arterials typically includes 15 to 25 per cent of the total roadway network and serves 40 to 60 per cent of total motor vehicle travel in the area.

Major Collectors and Minor Collectors

Collectors are the connectors between Arterials and Local Streets that serve to collect traffic and distribute it to the Arterial network. Collectors also serve to provide direct access to a wide variety of residential, commercial, and other land uses, and their design involves site-specific considerations. They provide direct service to neighborhoods and other local areas, and may border or traverse neighborhood boundaries. Parking is generally permitted on-street in residential areas. The distinction between Major Collectors and Minor Collectors is a matter of relative degree and includes consideration of functional role, size and character of areas served, traffic volume, and travel speed. Major Collectors connect Minor Collectors to the Arterial system.

Since Collectors are used for short distance trips between Local Streets and Arterials, they should be continuous in the spaces between Arterials. Collectors may also extend across Arterials, especially Major Collectors. To provide efficient traffic circulation and preserve amenities of neighborhoods, Collectors should desirably be spaced at about one-quarter to one-half mile intervals. Subdivision street layout plans should include Collectors as well as Local Streets in order to provide efficient traffic access and circulation.

Since Collectors generally carry higher traffic volumes than Local Streets, they may require a wider roadway cross section or added lanes at intersections with Arterials, to provide adequate capacity for both through traffic and turning movements. Operating speeds for Collectors are typically about 30 to 35 miles per hour. Since speeds are slower and more turn movements are expected, a higher speed differential and much closer intersection/access spacing can be used than on Arterials. On-street parking is permissible in residential areas. Direct access to abutting land is essential; parking and traffic controls may be necessary for safe and efficient through movement of moderate to low traffic volumes at key intersections. Collectors typically make up about 5 to 10 per cent of the total street system.

Collectors serve an important role in collecting and distributing traffic between Principal/Minor Arterials and Local Streets. Their identification is essential in planning and managing traffic ingress/egress and movement within residential neighborhoods as well as commercial and industrial areas. Existing Major Collectors are delineated on the Long Range Thoroughfare Plan, and planned new Major Collectors are shown as general alignments that should be considered and incorporated in subdivision platting and development planning. Alignments for existing and planned Minor Collectors are not shown on the plan, but should be appropriately considered and included in planning and plat review for subdivisions and development projects as needed.

Local Streets

Local Streets include all other streets and roads that are not included in higher classes. They include internal and access streets that allow direct access to residential and commercial properties and similar traffic destinations. Direct access to abutting land is their primary role, for all traffic originates from or is destined to abutting land. Through traffic and excessive speeds should be discouraged by using appropriate geometric designs, traffic control devices, curvilinear alignments, and discontinuous streets. On-street parking is generally permitted. Trip lengths on Local Streets are short, volumes are low, and speeds are slow, typically 20 to 30 miles per hour. Local Streets typically comprise about 65 to 80 per cent of the total street system in urban areas.

EXISTING TRANSPORTATION SYSTEM

Development of the Transportation Element for the Comprehensive Plan included analysis and evaluation of the existing transportation system. The roadway and traffic conditions of the highway and street network were identified and analyzed to assist in determining long-range needs for thoroughfare system development. Physical conditions of the roadway system and characteristics of existing travel patterns were identified based upon available information obtained through the City of Harlingen, Cameron County, Harlingen-San Benito Metropolitan Planning Organization (MPO), Texas Department of Transportation (TxDOT), and other governmental agencies. Existing transportation facilities and services for the Harlingen area are described in the following sections.

EXISTING ROADWAY CHARACTERISTICS CHARACTERISTICS

Characteristics of the existing roadway network include existing Federal and State Highways, traffic volumes, number of travel lanes, roadway surface types, and traffic control devices.

Major Highways

Federal and State Highways: Many thoroughfares in the study area have Federal (US), State Highway (SH), and Farm to Market (FM) highway designations. Harlingen area highways that are included as part of the National Highway System (NHS) include US 77, US 83, US 77/83, Loop 499, and Business 77 south of Loop 499. These highways connect the Harlingen area to the State and National Highway Systems that serve mobility on statewide, regional, national, and international scales. Improved connectivity with the National Highway System is vital for continued growth of the Valley's economy and international trade.

Interstate Highway System Designation

Recently, significant commitments have been made that will impact connectivity of the Valley with other regions of the United States and Mexico. There is a planned extension of the Interstate Highway System from the US

59 corridor from Houston to Laredo as an extension of I-69 south from Shreveport, with an alternate routing from Victoria south along US 77 to Harlingen, Brownsville and McAllen. The US 77 corridor from Victoria to Harlingen is the most direct connector between the Valley and I-10, and has great potential for further development to serve future traffic growth and international trade between the U.S., Mexico, and Canada.

Federal and State Highway Systems

Existing major highway routes in the Harlingen area are described in the following paragraphs and depicted on the Planning Area Map shown previously in Figure.

- *US Highway 77*: US 77 extends through the Harlingen area in a north-south direction and connects the area to Brownsville and the Texas/Mexico border to the south, and to Interstate Highway 37 at Corpus Christi and US 59 at Victoria to the north. US 77 joins with US 83 in Harlingen forming US 77/83, which extends south to Brownsville. US 77 is a four-lane divided freeway, with continuous two-lane frontage roads along the portion within the Harlingen area.
- *US Highway 83*: US 83 connects the Harlingen area to McAllen and other communities in the western part of the Lower Rio Grande Valley, and continues northwest upriver to Interstate Highway 35 and US 59 at Laredo. US 83 joins with US 77 in Harlingen, forming US 77/83, which extends south to Brownsville and the Texas/Mexico border. US 83 is a four-lane access controlled freeway with continuous two-lane frontage roads.
- *Business US 77 (Sunshine Strip)*: Business 77 parallels US 77 on the east and connects to US 77 northwest of Harlingen and to US 77/83 south of San Benito. Business 77 links the central business districts of Harlingen and San Benito. Between US 77 and FM 507, Business 77 is a five-lane roadway (with a continuous center two-way left-turn lane). South of FM 507 to Ed Carey Drive, Business 77 is a seven-lane facility (with a continuous center two-way left-turn lane). South of Ed Carey Drive, Business 77 converts back to a five-lane roadway (with a continuous center two-way left-turn lane).
- *Farm Road 106/206 (Tyler/Harrison Avenues)*: This facility is a one-way couplet extending in an east-west direction across the Harlingen area from the US 77/83 interchange to east of Business 77. Three lanes are provided in each direction, with Harrison Avenue (FM 106) serving the westbound direction and Tyler Avenue (FM 206) serving the eastbound direction. East of Business 77 at 27th Street, Tyler merges with Harrison and FM 106 continues eastward as a two-way, five lane facility (four lanes with a continuous center two-way left-turn lane) to Loop 499. East of Loop 499, FM 106 is a two

lane facility. East of FM 1595, FM 106 turns north and continues to Rio Hondo, where it merges with FM 508 and continues eastward to the Laguna Atascosa National Wildlife Refuge.

- *Farm Road 507 (Morgan Blvd./25th St.)*: This facility extends in a north-south direction along Morgan Boulevard and 25th Street from Business 77 to FM 508. Morgan Boulevard is a four-lane divided roadway and 25th Street is a two-lane roadway.
- *Farm Road 508 (Combes - Rio Hondo Road)*: This facility extends across the northern portion of the Harlingen area, from State Highway 107 at US 77 on the west, to FM 106 at Rio Hondo on the east. FM 508 is a two-lane roadway.
- *Farm Road 509*: FM 509 (Helen Moore Road) extends southwest from US 77/83 to FM 800 (Jim Bowie Road) and northeast to FM 1595 (Rio Hondo Road). This facility is a two-lane roadway. South of the Resaca de Los Fresnos, extends south to US 281 and the Free Trade Bridge at Los Indios. FM 509 is being developed as a planned "outer loop" which will extend along the east and north sides of the Harlingen area, from US 281 and the Los Indios Free Trade Bridge on the south, to US 77 on the north. Future extension of FM 509 is planned north from FM 1591 to connect with US 77.
- *Farm Road 800 (Bass Boulevard/Jim Bowie Road)*: FM 800 extends north and south from its intersection with US 77 across US 83 to the Arroyo Colorado and extends east to FM 2520. The western portion is named Bass Boulevard. The portion east of FM 1479 is named Jim Bowie Road. FM 800 is a two-lane roadway.
- *Farm Road 801 (Nixon Avenue)*: FM 801 is a continuation of Ed Carey Drive and extends southward from US 77/83 to FM 800. FM 801, also named Nixon Avenue, is a two-lane roadway.
- *Farm Road 1479 (Rangerville Road)*: FM 1479 extends from US 77/83 to US 281 through Rangerville and connects with FM 800 (Jim Bowie Rd.). The roadway is a two-lane facility except near US 77/83 where it expands to four and five lanes.
- *Farm Road 1595*: FM 1595 extends north from FM 106 to Rio Hondo Road, along the future alignment of Loop 509, and east along the Harlingen City Limits to SH 106. The roadway is a two-lane facility.
- *Loop 499 (Ed Carey Drive/Al Conway Drive)*: This facility forms a loop along the north and east sides of the City of Harlingen beginning at US 77 on the north, and extending east to the Airport and TSTC campus, continuing south to US 77/83, and connecting with FM 801 (Nixon Road) on the south. Loop 499 begins as a four-lane divided roadway at US 77 on the north, narrows to two-lanes, and then transitions to a five/four lane roadway at FM 507, intersects with FM 106 and Business 77, and continues to US 77/83. The portion of

Loop 499 from US 77 to FM 106 is named Al Conway Drive, and the portion from FM 106 to US 77/83 is named Ed Carey Drive. Ed Carey Drive is a five-lane roadway (with a continuous center two-way left turn lane).

- *Spur 54*: This facility connects US 77 and US 83, located just north of the US 77/83 interchange in Harlingen. The four-lane divided roadway provides an at grade connector which functions as part of the US 77/83 interchange, serving traffic movements for southbound US 77 to westbound US 83 and eastbound US 83 to northbound US 77.

Daily Traffic Volumes

Existing Traffic Volumes: Traffic volumes identify existing travel patterns and assist in determining the transportation system's ability to serve area travel demands. The identification of existing travel patterns and travel demands was based on available daily traffic volume counts conducted in the area. The source of traffic volumes is from volumes available from the Texas Department of Transportation (TxDOT) from 1994 traffic saturation counts.

Existing traffic conditions indicate that the following roadway segments have higher travel demands compared to other roads in the Harlingen area:

- Expressways US 83, US 77, and US 77/83 (up to 72,100 VPD);
- FM 106/206 (Tyler/Harrison Avenues), between US 77/83 and Loop 499 (14,300 to 41,600 VPD);
- Business 77 (Sunshine Strip), between US 77 and Loop 499 (9,700 to 29,000 VPD);
- Loop 499 (Ed Carey Drive), between US 77/83 and FM 106 (18,300 to 28,300 VPD);
- Dixieland Road, between US 77/83 and Lincoln Avenue (5,300 to 16,800 VPD);
- FM 507 (Morgan Boulevard), between Sunshine Strip and 25th Street (7,700 to 11,100 VPD);.
- Loop 499 (Al Conway Drive), between US 77 and FM 106 (16,000 to 51,000 VPD);
- Commerce Street, between Business 77 North and Business 77 South (5,200 to 15,300 VPD);
- M Street, between US 77/83 and Tyler Avenue (5,100 to 8,000 VPD); and,
- Lincoln Avenue, between US 83 and Dixieland Road (5,100 to 12,600 VPD).

Number of Lanes

Roadway Travel Lanes: The principal factor that determines the traffic carrying capacity of a roadway is the number of travel lanes available for moving traffic. The Harlingen area roadway network primarily consists of two-lane roadways and four-lane roadways. Five-lane roadway sections (which include a continuous center two-way left-turn lane) exist on Business 77, Loop 499, Commerce Street, FM 1479, Lincoln Avenue, Ed Carey Boulevard, and 13th Street. Seven-lane roadway sections (which include a continuous left-turn lane) exist on Business 77 between Treasure Hills Boulevard and Morgan Boulevard. US Highways that are four-lane divided roadways include US 77 and US 83. Three-lane roadway segments (which are one-way or include a continuous center two-way left-turn lane) exist along Harrison, Tyler, Van Buren, Monroe, Madison, and Helen Moore Road.

Type of Pavement

Roadway Surface: The type of surface on roadways has significant impact on the traffic capacity and utilization of the facilities. Surface types are generally categorized as either paved or unpaved. Most of the roadways within the Harlingen area are paved with concrete or asphalt surfaces, including all of the State Highways and Farm to Market Roads, and the majority of County Roads and City streets, which carry significant amounts of traffic in the urban and rural areas. The unpaved facilities are primarily rural roads that do not serve significant traffic volumes.

While mobility on the network is important, so it maintenance of the existing network. Without proper maintenance of the system mobility can be decreased. Users of the system will find alternate routes, some of which are not designed for increased volume, if the system is not maintained in good pavement condition with minimum cracks and potholes. Harlingen has an on-going resurfacing/rehabilitation program. City streets are surveyed and prioritized on a yearly basis. Approximately $500,000 is spent in resealing about 10 miles of existing roadways annually.

Traffic Control Devices

Traffic Signals, Signs and Markings: The principal means of facilitating safe and efficient traffic flow on the roadway network is through the application of traffic control devices including traffic signals, traffic signs and pavement markings. Of these, traffic signals have the greatest impact on traffic flow and roadway capacity.

In 2000, there were a total of 76 signalized intersections within the City of Harlingen with number 77 currently being installed on Dixieland Road in front of the Valle Vista Mall and Luby's. There is an interconnected traffic signal system in the central area of Harlingen. Most traffic signals located in the study area are time based or fixed-time, however there are some traffic actuated signals. Traffic actuated controllers generally provide greater flexibility in accommodating traffic demands by responding to the actual presence of vehicles at the intersection. With pretimed-timed controllers, the cycle length, timing, and phasing of traffic signals are of a fixed duration during specified periods of the day.

FREE TRADE BRIDGE AT LOS INDIOS

The Free Trade Bridge at Los Indios is the international border crossing and port of entry located approximately 8 miles south of Harlingen. It is owned by the Cities of Harlingen and San Benito, and Cameron County. Cameron County is responsible for the operation of the bridge. The U.S. border station and federal inspection facility is owned by the General Services Administration and operated by the U.S. Customs Service, Immigration and Naturalization Service (INS), and U.S. Department of Agriculture Animal and Plant Health Inspection Service (APHIS). Opened in November 1992, the Free Trade Bridge is one of the newest and most modern, industrial and commercial crossings along the U.S./Mexico border. Located on a 127 acre tract of land south of the intersection of FM 509 and US 281, the bridge facilities include bridge structure, border station, import and export inspection areas, impound lot, toll plaza, parking areas, and future development areas for uses such as customs brokers, bonded warehouses, duty free stores, money exchange houses, restaurants, and parking. The federal inspection station has the capacity to inspect up to 75 trucks per hour.

Traffic has increased steadily since opening and reached an annual total of 565,132 passenger and 73,904 trucks in fiscal year 1999 (October 1 to September 30). The Free Trade Bridge at Los Indios is significantly contributing to growth of international trade in the Harlingen area. The current use of the land area in the immediate vicinity of the bridge and along the FM 509 corridor between US 77/83 and US 281 is primarily agricultural. The corridor has significant potential for future industrial and commercial development. Carter Burgess conducted a Land Use Plan/Master Plan of the Free Trade Bridge area. The Plan outlines the potential land uses along FM509 as well as outlining future infrastructure needs for water, sewer, drainage and road improvements.

MagneTek has been operating a facility at the Free Trade Bridge since early 1999. MagneTek daily generates approximately 80 trucks southbound and northbound at the bridge. Highland Corporation is in the process of building a facility near the Free Trade Bridge. Fast processing times at the bridge allow companies to cross products and materials through the Free Trade Bridge in an efficient manner. In addition, the Cities of Harlingen, San Benito, and Los Indios have entered into a tri-regional Enterprise Zone which allows incentives to be offered to businesses desiring to locate in the area adjacent to the bridge and along the FM509 and FM1479 corridors extending into the city limits of Harlingen. (Both Harlingen Enterprise Zones are illustrated in Chapter 2, Demography and Economics with text on the effects of such designations.) Because of this, more development is anticipated in the near future.

Rio Grande Valley International Airport

The Rio Grande Valley International Airport serves the Harlingen area for air transportation access. Approximately 55 per cent of all air passenger traffic in the Lower Rio Grande Valley are served at Valley International Airport. Served by more than 40 daily flights on Southwest, Continental and American Airlines, it is the dominant commercial service airport for the Valley. Located on the northeast side of the City of Harlingen, Valley International Airport provides air transport service for the entire Lower Rio Grande Valley. Air transport service for passengers and cargo movement via Valley International Airport is a key asset for the City of Harlingen and the Valley.

Valley International Airport is owned and operated by the City of Harlingen Airport Board. The airport includes a total of 2,406 acres of land and consists of four runways and twelve taxiways. The primary runways consist of 8,000 and 7,250 foot long runways designed to serve commercial air carriers and commuters, as well as military and general aviation aircraft. Two secondary runways of 5,950 feet and 5,745 feet lengths also serve general aviation aircraft. All four runways are 150 feet wide. Other airport facilities include aircraft parking aprons, hangars, maintenance centers, manufacturing facilities, access roadways and parking facilities, a passenger terminal complex with 155,000 square feet of usable area, a 20,000 square foot Federal Inspection Services Building that accommodates personnel from 4 federal agencies, and an air cargo facility. Located on the east side of the airport is the Harlingen Industrial Airpark, which includes a Foreign Trade Zone.

Ground access to the Valley International Airport is provided by FM 507, FM 508, Loop 499, FM 509 (presently being extended to the Free Trade Bridge at Los Indios), US 77, and US 83. This roadway network connects the airport with the Harlingen area as well as the Rio Grande Valley in a direct and efficient manner by providing excellent roadway facilities and direct routes. Rail access for areas near the airport is provided by the Southern Pacific and Union Pacific Railroads. The Southern Pacific Railroad passes approximately two miles south of the airport property and an abandoned rail spur extends north along Loop 499 to the TSTC campus.

The development and operation of Valley International Airport is a key element in the economic development of the Harlingen area and, according to the Valley International Airport Master Plan, future aviation activity is forecast to increase significantly in commercial air carrier operations, passengers, air cargo, and general aviation activities over the next 20 years. In order to accommodate the expected growth, the Airport Master Plan recommends that the airport facilities be expanded, including the following planned improvements:

- Runway length;
- Passenger parking;
- Air cargo ramp and cargo Building;

- Additional navigational aids;
- General aviation aprons, T-hangars andcorporate/executive hangars; and
- Acquisition of property for clear zones and safety areas.

The development of Valley International Airport is greatly enhanced by its proximity to the Port of Harlingen and the Harlingen Industrial Parks, creating the potential of developing a multimodal transportation center by integrating air, maritime, rail and highway transportation, with convenient access to industrial plants and markets on both sides of the Texas/Mexico border. Efforts should be made in the future to improve direct access routes to the airport from North Expressway 77 and Loop 499. This will decrease surface travel times and be attractive to commercial passengers.

Railroads

The Harlingen area is served by two Class 1 railroads, the Southern Pacific and the Union Pacific Railroad companies. The Union Pacific Railroad crosses the area in a north-south direction paralleling Business 83 and bisecting the downtown area of Harlingen. The Southern Pacific Railroad extends in an east-west direction through the City of Harlingen serving the Port of Harlingen and Harlingen Industrial Parks. There are two switching yards in the study area. The Union Pacific Railroad (UPRR) switching yard, which is the major yard in the Rio Grande Valley south of Laredo, is located north and west of the Harlingen Central Business District, between Fair Park Boulevard and Wilson Road. The second switching yard, located north of Business 83 near Brazil Road, is operated by the Rio Valley Switching Company (RVSC). Construction is currently underway to move the switching yard to a location in Olmito.

The alignment of existing railways through the center of the urban area, and in particular the location of the rail switching yards, results in sever conflicts between railway and roadway traffic. Traffic movement and traffic safety considerations are important concerns relating to existing railroad and roadway at-grade crossings. The Brownsville Railroad Relocation Project to Olmito is designed to reduce the congestion at the Harlingen switching yard. Initially discussed in the early 1970's, this project, now under construction, will relocate most of the switching now done in Harlingen to the Olmito switching yard and significantly reduce the congestion in the Harlingen central business district. The Harlingen yard will continue to be utilized, but switching operations will become minimal.

Rerouting the railroad lines around Harlingen might also be considered to better serve the Industrial Parks and reduce traffic conflicts within the urbanized area. Railroad realignment along the east side of Harlingen would benefit the City by providing more direct rail access to the Industrial Parks, Industrial Airpark and Port of Harlingen, while removing traffic conflicts at existing grade crossings within the downtown area. The railroad companies would benefit from increased train operating speeds, reduced operating costs, and improved rail safety, as well as ability to accommodate expanded switching activity in larger and more efficient yard facilities. The railroads have had a major influence on the growth and development of the Valley and continue to perform important transportation service contributing to the area's economic vitality based on agriculture, manufacturing, and international trade.

Port of Harlingen and Harlingen Barge Channel

The Port of Harlingen is a 150-acre seaport terminal located on the east side of the Harlingen area, at the turning basin of the Harlingen Barge Channel on the Arroyo Colorado. The Harlingen Barge Channel extends 25 miles inland from the Gulf Intracoastal Waterway (GIWW) at Laguna Madre, to a turning basin on the Arroyo Colorado east of Harlingen. This waterway potentially connects Harlingen to Florida, Chicago and the Great Lakes. Constructed and maintained by the U.S. Army Corps of Engineers, the channel is 150 feet wide and 12 feet deep, and serves barges, coasters and small vessel traffic. The Port of Harlingen is the closest seaport to the Free Trade Bridge at Los Indios and is within a few miles of Valley International Airport and the Harlingen Industrial Parks.

The port accommodates barges and coasters for shipment of bulk commodities such as petroleum products, sugar, molasses, sand and gravel, liquid fertilizer, and cotton. Existing port facilities include a 650 foot long concrete general cargo wharf, 100 foot long dry bulk cargo wharf, five smaller docks, and covered storage facilities. Railroad service is provided by the Southern Pacific Railroad. The port can be greatly enhanced by continued expansion and improvement of its facilities and access needs. Future planned improvements include dredging the harbor to widen the turning basin and to add breasting structures on which to tie barges.

In 1999, the Port of Harlingen handled approximately 850,000 tons of cargo. Primary import commodities included petroleum, steel, sand, gravel, cement, agricultural chemicals and fertilizers. Primary exports are raw sugar, grain, cotton and other bulk products. The Port is part of Foreign Trade Zone #62.

Pedestrian and Bicycle Facilities

The City of Harlingen has many wide streets, railroad corridors, irrigation and drainage canals, and parks and recreational areas that represent opportunities for future development of bicycle and pedestrian facilities. These opportunities can be incorporated as transportation enhancement projects such as bikeways, pedestrian facilities, historic sites, and scenic beautification areas.

Arroyo Colorado Hike and Bike Trail

Implementation of a 2.1-mile "Bike and Jog Trail" along the Arroyo Colorado was funded as a Transportation Enhancement project by the Texas Department of Transportation in 1994. Groundbreaking took place in March, 2000. This Arroyo Colorado Hike and Bike Trail will provide an alternative mode of transportation and serve recreational needs for area residents. Plans for future phases will eventually create a 4.1 mile trail connecting five City Parks. Bikeway Master Plan A Bicycle Facilities Master Plan for the Harlingen-San Benito Metropolitan Area was prepared as part of the Metropolitan Transportation Plan completed in 1994. There are planned bicycle facilities for the Harlingen area, including on-street bikeways and off-street bike trails. Implementation of recommended bicycle and pedestrian improvements will enhance the overall surface transportation system for the area.

Pedestrian Improvements

Pedestrian walkways, sidewalks and crosswalks are part of the City's existing transportation system that serves the need for pedestrian movement in residential neighborhoods, commercial business areas, and around schools, parks and other community facilities. Pedestrian facilities are needed in older areas of the community that developed before walkways were required by the City's development ordinances, as well as in newly developed areas. Pedestrian and bicycle facilities should be constructed in compliance with requirements of the Americans with Disabilities Act (ADA).

Intercity Bus Service

Intercity bus transportation service for passengers and packages is provided by Valley Transit Company and Greyhound Bus Lines. In addition to intercity bus service, airport limousine passenger shuttle service between Valley International Airport, South Padre Island and other Valley destinations is operated by SURFTRAN and Gray Line Tours. Although intercity bus travel has declined due to the perceived convenience and economical cost characteristics of automobile and airline travel, the need for intercity bus service remains important for many residents of the Lower Rio Grande Valley and northern Mexico. The intercity bus terminal for both bus companies is located in downtown Harlingen. Future consideration should be given to the development of an intermodal transportation terminal incorporating intercity buses, taxis, airport shuttles, and auto parking in a single location, and extension of intercity bus service to cover Valley International Airport, Texas State Technical College, and other areas that have the demand for this type of service.

Public Transit

The Harlingen area started the Harlingen-San Benito Express on January

10, 2000. This system consists of two, eleven passenger vans, wheelchair accessible. The system runs Monday through Friday, 8 a.m. - 5 p.m. and is a demand response system meaning that the vans will pick up and drop off at any location within the City limits of Harlingen and San Benito with a 24 hour advance call request. The system is funded through Federal and State transit funds with a local match provided by the two cities. The system is primarily for the elderly and the handicapped, but will accommodate others as space permits.

This public transit service provides increased mobility and access to social services, health care, and community activities for residents who are transportation disadvantaged, including the elderly, handicapped, low income persons, and youths. Winter Texans also represent a significant population of potential transit users who would benefit from increased mobility provided by a public transit system. Other available public transportation is limited to privately owned taxis, airport limousine shuttle service by SURFTRAN and Gray Line Tours, school buses, and special transportation service provided by agencies such as Amigos del Valle and other social service and religious organizations.

Major Traffic Generators

The location and character of land uses that generate large numbers of trips have a major influence on traffic volumes and flow patterns. Major traffic generators were identified and considered in reviewing the transportation system and developing the Transportation Element.

Major traffic generators in the Harlingen area include the following uses and activities:

- Downtown Harlingen;
- Valley Baptist Medical Center;
- Rio Grande Valley State Center for MHMR;
- South Texas State Hospital;
- Valle Vista Mall and other nearby retail uses;
- Sun Valley Mall;- Treasure Hills Shopping Center;
- Rio Grande Valley International Airport;
- Texas State Technical College;
- Valley Greyhound Racetrack (re-opened March 2000);
- J. J. Victor Park;
- Harlingen Sports Complex;
- Lon C. Hill Park, Municipal Auditorium, Baseball Stadium;
- Pendleton Park;
- Harlingen Industrial Parks I - IV;
- Port of Harlingen;- Harlingen High School and Harlingen High School South;
- Marine Military Academy;

- Valley Baptist Academy; and,
- Other existing and future large employers and activity centers.

Harlingen's Long Range Thoroughfare Plan

The Transportation Element includes the Long Range Thoroughfare Plan for the City of Harlingen, which identifies the existing and proposed thoroughfare system of Expressways, Arterials and Collector streets. Harlingen's thoroughfare system is comprised of existing and planned freeways, expressways, and major streets and highways, which require wider or new rights-of-way and may ultimately be developed as two-lane or multi-lane roadways with various cross sections.

Purpose of Thoroughfare Planning

The Long Range Thoroughfare Plan is the City of Harlingen's general plan for guiding thoroughfare system development, including the planned widening and extension of its roads, streets, and public highways within the municipality and its Extraterritorial Jurisdiction. The plan indicates the needed rights-of-way, general alignments, and typical sections for planned new roadways, as well as for widening and extensions of existing thoroughfares. Proposed alignments are shown for planned new roadways and roadway extensions, and the actual alignments may vary depending upon future development. The Long Range Thoroughfare Plan shows approximate alignments and right-of-way requirements for planned thoroughfares that should be considered in platting of subdivisions, right-of-way dedication, and construction of major roadways. The plan does not show future alignments for new Local Streets, because these streets function principally to provide access to adjacent land and their future alignments may vary depending upon specific development plans. Collector and Local Street alignments should be determined by the City and developers as part of planning for new development.

Benefits of Thoroughfare Planning

The primary objective of the Long Range Thoroughfare Plan is to ensure the reservation of adequate rights-of-way on appropriate alignments and of sufficient width to allow the orderly and efficient expansion and improvement of the thoroughfare system to serve existing and future transportation needs.

The benefits provided by effective thoroughfare planning and implementation include the following objectives:

- Reservation of adequate rights-of-way for future long-range transportation improvements;
- Making efficient use of available resources by designating and recognizing the major streets that will likely require higher design of improvements;

- Minimizing the amount of land required for street and highway purposes;
- Identifying the functional role that each street should be designed to serve in order to promote and maintain the stability of traffic flow and land use patterns;
- Informing citizens of the streets that are intended to be developed as arterial and collector thoroughfares, so that private land use decisions can anticipate which streets will become major traffic facilities in the future;
- Providing information on thoroughfare improvement needs which can be used to determine priorities and schedules in the City's capital improvement program and capital budget; and,
- Minimizing the negative impacts of street widening and construction on neighborhood areas and the overall community, by recognizing where future improvements may be needed and incorporating thoroughfare needs in the city's comprehensive planning process.

Implementing the Major Street and Highway Plan

Implementation of thoroughfare system improvements occurs in stages over time as the City grows and, over many years, builds toward the ultimate thoroughfare system shown in the Major Street and Highway Plan. The fact that a planned thoroughfare is shown in the plan does not represent a commitment to a specific time frame for construction, nor that the City will build the roadway improvement. Individual thoroughfare improvements may be constructed by a variety of implementing agencies including the City of Harlingen, Cameron County, and Texas Department of Transportation (TxDOT), as well as private developers and land owners for sections of roadways located within or adjacent to their property.

The City of Harlingen, Cameron County, and TxDOT, as well as residents, land owners and developers, can utilize the Long Range Thoroughfare Plan in making decisions relating to the planning, coordination and programming of future development and transportation improvements. Review by the City of preliminary and final plats for proposed subdivisions in accordance with the Subdivision Ordinance should include consideration of compliance with the Long Range Thoroughfare Plan, in order to ensure consistency and availability of sufficient right-of-way for the general roadway alignments shown in the plan. By identifying thoroughfare locations where right-of-way is needed, land owners and developers can consider the roadways in their subdivision planning, dedication of public right-of-way, and provision of set backs for new buildings, utility lines, and other improvements located along the rights-of-way for existing and planned thoroughfares.

The Long Range Thoroughfare Plan will have long-reaching effects on the growth and development of the Harlingen area, since it guides the reservation of right-of-way needed for future thoroughfare improvements. The plan has important influence on the pattern of movement and the desirability of areas as locations for development and land use. While other elements of the Comprehensive Plan look at foreseeable changes and needs over a 20 to 25-year period, thoroughfare planning requires an even longer-range perspective extending into the very long-term future. For example, consider the enduring impact of Lon C. Hill's original 1911 Plat Plan for Harlingen, which established the City's street system that, 85 years later, still serves the central core of Harlingen today, with 80 foot wide avenues running east-west and 60 foot wide streets running north-south. Future changes in transportation technology, cost structure, service demands for the transportation system, and resulting long-term shifts in urban growth and development patterns require a farsighted and visionary approach to thoroughfare planning decisions.

Thoroughfare Development Requirements and Standards

Planning, design and construction of thoroughfares must comply with the development standards that are contained in the City of Harlingen's Subdivision Ordinance. Requirements for thoroughfare development include standards and criteria governing the following characteristics of thoroughfares:

- *Location and Alignment of Thoroughfares*: The general location and alignment of thoroughfares must be in conformance with the Major Street and Highway Plan contained in the Comprehensive Plan. Subdivision plats should provide for dedication of needed right-of-way for thoroughfares within or bordering the subdivision.
  Any major changes in thoroughfare alignment that are inconsistent with the plan require the approval of the Planning and Zoning Commission and City Commission through a public hearing process. A major change would include any proposal that involves the addition or deletion of established thoroughfare designations, or changes in the planned general alignment of thoroughfares that would affect parcels of land beyond the specific tract in question.
- *Right-of-Way and Pavement Width*: The pavement width and right-of-way width for thoroughfares and other public streets should conform to the following City minimum standards, unless a variance is granted:
  - *Major Arterials*: 80 feet pavement width within 120 feet right-of-way width;
  - *Minor Arterials*: 60 feet pavement width within 100 feet right-of-way width;
  - *Major Collectors*: 48 feet pavement width within 70 feet right-of-way width;
  - *Minor Collectors*: 42 feet pavement width within 60 feet right of way width; and,

- *Local Streets*: 37 feet pavement width within 50 feet right-of-way width.

Plats that include or are bordered by an existing thoroughfare with insufficient right-of-way width should be required to dedicate land to compensate for any right-of-way deficiency of that thoroughfare. When a new thoroughfare extension is proposed to connect with an existing thoroughfare that has narrower right-of-way, a transitional area should be provided.

- *Continuation and Projection of Streets:* Existing streets in adjacent areas should be continued and, when an adjacent area is undeveloped, the street layout should provide for future projection and continuation of streets into the undeveloped area. Stub streets not exceeding 200 feet in length may be provided to permit future expansion of the street system
- *Location of Street Intersections*: New intersections of subdivision streets with existing thoroughfares within or bordering the subdivision should be planned to align with existing intersections, where feasible, to avoid creation of off-set or "jogged" intersections and to provide for continuity of existing streets, especially collectors and higher classes of thoroughfares.
- *Angle of Intersection*: The angle of intersection for street intersections should be as nearly at a right angle as possible. Corner cutbacks or radii should be required at the acute corner of the right-of-way line, to provide adequate sight distance at intersections.
- *Off-Set Intersections*: Offset or "jogged" street intersections must have a minimum separation of 125 feet between the centerlines of the intersecting streets.
- *Cul-De-Sacs*: Cul-de-sac streets should have a maximum length of no more than 600 feet, with a paved turnaround pad of 100 feet diameter.
- *Residential Lots Fronting on Arterials*: Wherever feasible, subdivision layout should avoid the creation of residential lots fronting on major or minor arterials, with direct driveway access to the arterial street. Lots should be accessed from collector or local streets within or bordering the subdivision.
- *Geometric Design Standards and Guidelines*: Other requirements and guidelines for the geometric design of thoroughfares and public streets are provided in the Subdivision Ordinance. Standard roadway cross sections for thoroughfares and other public streets are contained in the geometric design guidelines.

Standard Roadway Cross Sections

While thoroughfare classification reflects the functions that roadways serve as part of the street and highway network, the cross section of a roadway is related to traffic volume, design capacity, and Level of Service. Standard roadway cross sections for the different classes of thoroughfares and Local Streets are shown in Figures. These cross sections identify the required minimum dimensional criteria for right-of-way and pavement width to comply with the City's Subdivision Ordinance. Typical configurations for the number of travel lanes, median, and parking are also shown. The classes of roadways correspond to the functional classifications shown in the Long Range Thoroughfare Plan.

In the administration and enforcement of the Long Range Thoroughfare Plan, special cases and unique situations will occasionally arise where existing physical conditions and development constraints in certain areas conflict with the need for widening of designated thoroughfares to the planned right-of-way width and roadway cross section. Such special circumstances require a degree of flexibility and adaptability in the administration and implementation of the plan. Acceptable minimum design criteria and special roadway cross sections may have to be applied in constrained areas where existing conditions limit the ability to meet desirable standards and guidelines. Special roadway cross sections should be determined on a case by case basis when a unique design is needed, and subject to approval by the Planning and Zoning Commission. The standard roadway cross sections should be used in all newly developing areas and, whenever possible, in existing developed areas.

Goals and Objectives

The goals and objectives for the Transportation Element of the Comprehensive Plan are outlined in this section. These goals and objectives are the framework for the Transportation Element.

2

Transportation and Economic Development

THE ECONOMIC IMPORTANCE OF TRANSPORTATION

Like many economic activities that are intensive in the use of infrastructures, the transport sector is an important component of the economy impacting on development and the welfare of populations. A relation between the quantity and quality of transport infrastructure and the level of economic development is apparent. When transport systems are efficient, they provide economic and social opportunities and benefits that result in positive multipliers effects such as better accessibility to markets, employment and additional investments. When transport systems are deficient in terms of capacity or reliability, they can have an economic cost such as reduced or missed opportunities and lower quality of life. Efficient transportation reduces costs, while inefficient transportation increases costs. The impacts of transportation are not always intended, and can have unforeseen or unintended consequences such as congestion. Transport also carries an important social and environmental load, which cannot be neglected.

The added value and employment effects of transport services usually extend beyond employment and added value generated by that activity; indirect effects are salient. For instance, transportation companies purchase a part of their inputs from local suppliers. The production of these inputs generates additional value-added and employment in the local economy. The suppliers in turn purchase goods and services from other local firms. There are further rounds of local re-spending which generate additional value-added and employment. Similarly, households that receive income from employment in transport activities spend some of their income on local goods and services. These purchases result in additional local jobs and added value. Some of the household income from these additional jobs is in turn spent on local goods and services, thereby creating further jobs and income for local households. As a result of these successive rounds of re-spending in the framework of local purchases, the overall impact on the economy exceeds the initial round of output, income and employment generated by passenger and freight transport activities.

Thus, from a general standpoint the economic impacts of transportation can be direct, indirect and related:

- Direct impacts (also known as induced) the outcome of accessibility changes where transport enables employment, added value, larger markets and enables to save time and costs.
- Indirect impacts the outcome of the economic multiplier effects where the price of commodities, goods or services drop and/or their variety increases. Indirect value-added and jobs are the result of local purchases by companies directly dependent upon transport activity. Transport activities are responsible for a wide range of indirect value-added and employment effects, through the linkages of transport with other economic sectors (*e.g.* office supply firms, equipment and parts suppliers, maintenance and repair services, insurance companies, consulting and other business services).
- Related impacts the outcome of economic activities and firms partly relying on efficient transport services for both passengers and freight. For instance, the steel industry requires cost efficient import of iron ore and coal for the blast furnaces and export activities for finished products such as steel booms and coils. Manufacturers and retail outlets and distribution centers handling imported containerized cargo rely on efficient transport and seaport operations.

Mobility is one of the most fundamental and important characteristics of economic activity as it satisfies the basic need of going from one location to the other, a need shared by passengers, freight and information. All economies and regions do not share the same level of mobility as most are in a different stage in their mobility transition towards motorized forms of transport. Economies that possess greater mobility are often those with better opportunities to develop than those with scarce mobility. Reduced mobility impedes development while greater mobility is a catalyst for development. Mobility is thus a reliable indicator of development. Providing this mobility is an industry that offers services to its customers, employs people and pays wages, invests capital and generates income.

The economic importance of the transportation industry can thus be assessed from a macroeconomic and microeconomic perspective:

- At the macroeconomic level (the importance of transportation for a whole economy), transportation and the mobility it confers are linked to a level of output, employment and income within a national economy. In many developed countries, transportation accounts between 6 per cent and 12 per cent of the GDP.
- At the microeconomic level (the importance of transportation for specific parts of the economy) transportation is linked to producer, consumer and production costs. The importance of specific transport activities and infrastructure can thus be assessed for eachsector of

the economy. Transportation accounts on average between 10 per cent and 15 per cent of household expenditures while it accounts around 4 per cent of the costs of each unit of output in manufacturing, but this figure varies greatly according to sub sectors.

Transportation links together the factors of production in a complex web of relationships between producers and consumers. The outcome is commonly a more efficient division of production by an exploitation of geographical comparative advantages, as well as the means to develop economies of scale and scope. The productivity of space, capital and labor is thus enhanced with the efficiency of distribution and personal mobility. It is acknowledged that economic growth is increasingly linked with transport developments, namely infrastructures but also managerial expertise is crucial for logistics.

The following categories of impacts can be assessed:

- *Core*: The most fundamental impacts of transportation relate to the physical capacity to convey passengers and goods and the associated lower costs. This involves the setting of routes enabling new or existing interactions between economic entities.
- *Operational*: Improvement in the time performance, notably in terms of reliability, as well as reduced loss or damage. This implies a better utilization level of existing transportation assets benefiting its users as passengers and freight are conveyed more rapidly and with less delays.
- *Geographical*: Access to a wider market base where economies of scale in production, distribution and consumption can be improved. Increases in productivity from the access to a larger and more diverse base of inputs (raw materials, parts, energy or labor) and broader markets for diverse outputs (intermediate and finished goods). Another important geographical impacts concerns the influence of transport on the location of activities.

TRANSPORTATION AND ECONOMIC OPPORTUNITIES

Transportation developments that have taken place since the beginning of the industrial revolution have been linked to growing economic opportunities. At each stage of human societal development, a particular transport mode has been developed or adapted. However, it has been observed that throughout history that no single transport has been solely responsible for economic growth. Instead, modes have been linked with the function and the geography in which growth was taking place. The first trade routes established a rudimentary system of distribution and transactions that would eventually be expanded by long distance maritime shipping networks and the setting of the first multinational corporations. Major flows of international migration that occurred since the 18th century were linked with the expansion of international and continental transport systems that radically shaped emerging economies such as in North America and Australia. Transport has played a catalytic role in these migrations, transforming the economic and social geography of many nations.

Transportation has been a tool of territorial control and exploitation, particularly during the colonial era where resource-based transport systems supported the extraction of commodities in the developing world and forwarded them to the industrializing nations of the time. More recently, port development, particularly container ports, has been of strategic interest as a tool of integration to the global economy as the case of China illustrates. There is commonly a direct relation between foreign trade and container port volumes.

Due to demographic pressures and increasing urbanization, developing countries are characterized by a mismatch between limited supply and growing demand for transport infrastructure. While some regions benefit from the development of transport systems, others are often marginalized by a set of conditions in which inadequate transportation plays a role. Transport by itself is not a sufficient condition for development. However, the lack of transport infrastructures can be seen as a constraining factor on development. In developing countries, the lack of transportation infrastructures and regulatory impediments are jointly impacting economic development by conferring higher transport costs, but also delays rendering supply chain management unreliable. A poor transport service level can negatively affect the competitiveness of regions and corporations and thus have a negative impact on the regional added value and employment. In 2007, the World Bank published its first ever report which ranked nations according to their logistics performance based on the so-called Logistics Performance Index. Investment in transport infrastructures is thus seen as a tool of regional development, particularly in developing countries and for the road sector.

The standard assumption is that transportation investments tend to be more wealth producing as opposed to wealth consuming investments such as services. Still, several transportation investments can be wealth consuming if they merely provide convenience, such as parking and side walks, or service a market size well below any possible economic return, with for instance projects labeled "bridges to nowhere". In such a context, transport investment projects can be counter productive by draining the resources of an economy instead creating wealth and additional opportunities. Efficient and sustainable transport markets and systems play a key role in regional development although the direction of causality between transport and wealth generation is not always clear. In a number of regions around the world, transport markets and related transport infrastructure networks are seen as key drivers in the promotion of a more balanced and sustainable development of the region or even the entire continent, particularly by improving accessibility and the situation of weaker regions and disadvantaged social groups.

There is also a tendency for transport investments to have declining marginal returns. While initial infrastructure investments tend to have a high return since they provide an entirely new range of mobility options, the more the system is developed the more likely additional investment would result in lower returns. At some point, the marginal returns can be close to zero or even negative, implying a shift of transport investments from wealth producing to wealth consuming. A common fallacy is assuming that additional transport investments will have a similar multiplying effect than the initial investments had, which can lead to capital misallocation.

The most common reasons for the declining marginal returns of transport investments are:

- *High levels of existing infrastructure*: In a context of high level of accessibility and transportation networks that are already extensive, further investments usually result in marginal improvements. This means that the economic impacts of transport investments tend to be significant when infrastructures were previously inexistent or deficient and marginal when an extensive network is already present. Additional investments can thus have limited impact outside convenience.
- *Economic changes*: As economies develop, the nature of their economies tends to shift from the primary (resource extraction) and secondary (manufacturing) sectors towards services. These sectors rely on different transport systems. While an economy depending on manufacturing will rely on road, rail and port infrastructures, a service economy is more oriented towards the efficiency of logistics and urban transportation. In all cases transport infrastructure are important
- *Economies of agglomeration*: Due to clustering and agglomeration, several locations develop advantages that cannot be readily reversed through improvements in accessibility. Transportation can be a factor of concentration and dispersion depending on the context. Less accessible regions thus do not necessarily benefit from transport investments if they are embedded in a system of unequal relations. Therefore, each development project must be considered independently.

TYPES OF TRANSPORT IMPACTS

The relationship between transportation and economic development is difficult to formally establish and has been debated for many years. In some circumstances transport investments appear to be a catalyst for economic growth while in others, economic growth puts pressures on existing transport infrastructures and incite additional investments. At start there are different impacts on the transport providers (transport companies) and the transport users. There are several layers of activity that transportation can valorize, from a suitable location that experiences the development of its accessibility through infrastructure investment to a better usage of existing transport assets through management.

This is further nuanced by the nature, scale and scope of possible impacts:

- Timing of the development varies as the impacts of transportation can precede, occur during or take place after economic development. The lag, concomitant and lead impacts make it difficult to separate the specific contributions of transport to development. Each case study appears to be specific to a set of timing circumstances that are difficult to replicate elsewhere.
- Types of impacts vary considerably. The spectrum of impacts range from the positive through the permissive to the negative. In some cases transportation impacts can promote, in others they may hinder economic development in a region. In many cases, few, if any, direct linkages could be clearly established.

Cycles of economic development provide a revealing conceptual perspective about how transport systems evolve in time and space as they include the timing and the nature of the transport impact on economic development.

This perspective underlines that after a phase of introduction and growth, a transport system will eventually reach a phase of maturity through geographical and market saturation. There is also the risk of overinvestment when economic growth is credit driven, which can lead to significant misallocations of capital, including in the transportation sector. The outcome is a surplus capacity in infrastructures and modes creating deflationary pressures that undermines profitability. In periods of recession that commonly follow periods of expansion, transportation activities may experiment a setback, namely in terms of lower demand and a scarcity of capital investment.

Transport, as a technology, typically follows a path of experimentation, introduction, adoption and diffusion and, finally, obsolescence, each of which has an impact on the rate of economic development. They follow a cyclic behavior where a high level of benefits and productivity is realized in the early phase while later phases are facing diminishing returns. Containerization is a relevant example of such a diffusion behavior. As most innovations are eventually abandoned, many technologies go through what can be called a "hype phase" with unrealistic expectations. In addition, transport modes and infrastructures are depreciating assets that constantly require maintenance and upgrades. At some point, their useful lifespan is exceeded and the vehicle must be retired or the infrastructure rebuilt. Thus, transport investments for their amortization must consider the lifespan of the concerned mode or infrastructure.

In general, transport technology can be linked to five major waves of economic development where a specific mode or system emerged:

- *Seaports*: Linked with the early stages of European expansion from the 16th to the 18th centuries. They supported the development of international trade through colonial empires, but were constrained by limited inland access.
- *Rivers and canals*: The first stage of the industrial revolution in the late 18th and early 19th centuries was linked to the development of canal systems in Western Europe and North America, mainly to transport heavy goods. This permitted the development of rudimentary and constrained inland distribution systems.
- *Railways*: The second stage of industrial revolution in the 19th century was intimately linked to the development and implementation of rail systems, some transcontinental, enabling a more flexible inland transportation system.
- *Roads*: The 20th century saw the development of road transportation systems and automobile manufacturing. Individual transportation became a commodity available to the masses, especially after the Second World War. This process was reinforced by the development of national highway systems.
- *Airways and information*: The later part of the 20th century saw the development of global air and telecommunication networks in conjunction with the globalization of economic activities. New organization, control and maintenance capacities were made possible. Electronic communications have become consistent with transport functions, especially in the rapidly developing realm of logistics and supply chain management.

TRANSPORT AS A FACTOR OF PRODUCTION

Contemporary trends have underlined that economic development has become less dependent on relations with the environment (resources) and more dependent on relations across space. While resources remain the foundation of economic activities, the commodification of the economy has been linked with higher levels of material flows of all kinds. Concomitantly, resources, capital and even labor have shown increasing levels of mobility.

This is particularly the case for multinational firms that can benefit from transport improvements in two significant markets:

- Commodity market. Improvement in the efficiency with which firms have access to raw materials and parts as well as to their respective customers. Thus, transportation expands opportunities to acquire and sell a variety of commodities necessary for industrial and manufacturing systems.
- Labor market. Improvement in the access to labor and a reduction

in access costs, mainly by improved commuting (local scale) or the use of lower cost labor (global scale).

A common fallacy in assessing the importance and impact of transportation on the economy is to focus only on transportation costs, which tend to be relatively low (5 to 10 per cent of the value of a good). Transportation is an economic factor of production of goods and services, implying that relatively small changes can have substantial impacts in on costs, locations and performance. An efficient transport system with modern infrastructures favors many economic changes, most of them positive. It provides market accessibility by linking producers and consumers.

The major impacts of transport on economic processes can be categorized as follows:

- *Geographic specialization*: Improvements in transportation and communication favor a process of geographical specialization that increases productivity and spatial interactions. An economic entity tends to produce goods and services with the most appropriate combination of capital, labor, and raw materials. A given area will thus tend to specialize in the production of goods and services for which it has the greatest advantages (or the least disadvantages) compared to other areas as long as appropriate transport is available for trade. Through geographic specialization supported by efficient transportation, economic productivity is promoted. This process is known in economic theory as comparative advantages.
- *Large scale production*: An efficient transport system offering cost, time and reliability advantages permits goods to be transported over longer distances. This facilitates mass production through economies of scale because larger markets can be accessed. The concept of "just-in-time" has further expanded the productivity of production and distribution with benefits such as lower inventory levels and better responses to shifting market conditions. Thus, the more efficient transportation becomes, the larger the markets that can be serviced and the larger the scale of production.
- *Increased competition*: When transport is efficient, the potential market for a given product (or service) increases, and so does competition. A wider array of goods and services becomes available to consumers through competition which tends to reduce costs and promote quality and innovation. Globalization has clearly been associated with a competitive environment that spans the world.
- *Increased land value*: Land which is adjacent or serviced by good transport services generally has greater value due to the utility it confers to many activities. In some cases, the opposite can be true if related to residential activities. Land located near airports and highways, near noise and pollution sources, will thus suffer from corresponding diminishing land value.

Transport also contributes to economic development through job creation and its derived economic activities. Accordingly, a large number of direct (freighters, managers, shippers) and indirect (insurance, finance, packaging, handling, travel agencies, transit operators) employment are associated with transport. Producers and consumers take economic decisions on products, markets, costs, location, prices which are themselves based on transport services, their availability, costs and capacity.

SOCIOECONOMIC IMPACTS

While many of the economic impacts of transportation are positive, there are also significant negative impacts that are assumed by individuals or by the society in one way or another.

Among the most significant are:

- *Mobility gaps*: Since mobility is one of the fundamental components of the economic benefits of transportation, its variations are likely to have substantial impacts on the opportunities of individuals. Mobility needs do not always coincide due to several factors, namely the lack of income, lack of time, lack of means and the lack of access. People's mobility and transport demands thus depend on their socioeconomic situation. The higher the income, the higher the mobility, which may give rise to substantial mobility gaps between different population groups. Gender gaps exist in mobility as women tend to have lower incomes. Mobility gaps are particularly prevalent for long distance travel. With the development of air transport, a segment of the global population has achieved a very high level of mobility for their business and leisure activities, while the great majority of the global population has little mobility. This issue is expected to become more acute as the population of many advanced economies is aging rapidly, which implies that access to mobility will not be an income issue but an age issue. By 2020, about 10 per cent of the global population (719 million) will be over 65 while by 2050 it will be 16 per cent (1,492 million).
- *Costs differences*: Locations that have low levels of accessibility, such as landlocked countries, tend to have higher costs for many goods (sometimes basic necessities such as food) as most have to be imported, often over long distances. The resulting higher transport costs inhibit the competitiveness of such locations and limits opportunities. Consumers and industries will pay higher prices, impacting on their welfare (disposable income) and competitiveness.
- *Congestion*: With the increased use of transport systems, it has become common for parts of the network to be used above design capacity. Congestion is the outcome of such a situation with its associated costs, delays and waste of energy. Distribution systems

that rely upon on-time deliveries are particularly susceptible to congestion.

- *Accidents*: The use of transport modes and infrastructure is never entirely safe. Every motorized vehicle contains an element of danger and nuisance. Due to human errors and various forms of physical failures (mechanical or infrastructural) injuries, damages and even death occur. Accidents tend to be proportional to the intensity of use of transport infrastructures which means the more traffic the higher the probability for an accident to occur. They have important socioeconomic impacts including healthcare, insurance, damage to property and the loss of life. The respective level of safety depends on the mode of transport and the speed at which an accident occurs. No mode is completely safe but the road remains the most dangerous medium for transportation, accounting for 90 per cent of all transport accidents on average. At the global level about 1.3 million people died in road accidents in 2010 in addition to 50 million injuries. China has one of the highest car accident death rates in the world, with more than 110,000 fatalities per year (300 per day), a factor mainly due to recent growth in vehicle ownership.

The emission of pollutants related to transport activities has a wide range of environmental consequences that have to be assumed by the society, more specifically on four elements:

- *Air quality*: Atmospheric emissions from pollutants produced transportation, especially by the internal combustion engine, are associated with air pollution and, arguably, global climate change. Some pollutants (NOx, CO, O3, VOC, etc.) can produce respiratory troubles and aggravate cardiovascular illnesses. In urban regions, about 50 per cent of all air pollution emanates from automobile traffic.
- *Noise*: A major irritant, noise can impact on human health and most often human welfare. Noise can be manifested in three levels depending on emissions intensity; psychological disturbances (perturbations, displeasure), functional disturbances (sleep disorders, loss of work productivity, speech interference) or physiological disturbances (health issues such as fatigue, and hearing damage). Noise and vibration associated with trains, trucks, and planes in the vicinity of airports are major irritants.
- *Water quality*: Accidental and nominal runoff of pollutants from transport such as oil spills, are sources of contamination for both surface water and groundwater.
- *Land take*: Transport is a large consumer of space when all of its supporting infrastructure and equipment are considered. Furthermore, the planning associated with these structures does not

always consider aesthetic values as is often the case in the construction of urban highways. These visual impacts have adverse consequences on the quality of life of nearby residents.

THE PUBLIC COSTS OF TRANSPORT INVESTMENTS AND OPERATIONS

To slightly misquote the late American political scientist, John Kenneth Galbraith, "The only function of traffic forecasting is to make astrology look respectable". While Galbraith was, of course, talking about economics, the sentiment extends to traffic forecasting. We are simply not very good at it, or in some instances we allow vested interests to manipulate predictions to achieve their own self-serving ends. The probable distortions in transport policy that result from this vary with context and the exact nature of where the errors or "interpretations" in forecasting lie. Using the term bias as shorthand for a portfolio of issues, if the forecasting bias is exactly or nearly the same across modes, space and time then this may well have little impact on the relative allocation of resources within transport. If, of the other hand the forecasts of the public costs of transport investments and operations are consistently low, or forecast demand consistently high, but accurate in other sectors of the economy, then transport will get an unjustified share of tax payers money.

One of the problems, however, is that it is difficult to know whether poor forecasting is the only, or even the main, problem in activities like transport investment appraisal. Other techniques, such as cost-benefit analysis, have their own limitations, but it seems wise as practical expedience to provide the best and most appropriate forecasts irrespective of other issues. Before looking at cures, or at least treatments for the problem, we offer a few comments on both the nature of the forecasting biases and what we know about the underlying causes. This provides a foundation for looking at remedial strategies.

THE SCALE OF THE PROBLEM AND ITS IMPACTS

The poor quality of traffic and other transport forecasts have been known for some time, and have periodically been highlighted in studies and in the light of experience with their use. The 1970s saw considerable debate in the UK, for example, over the inaccuracies in the forecasts of car ownership, a major input into traffic modelling, and of the traffic forecasts themselves in the 1980s. The forecasts for the M25 London orbital road, for instance, were that on 21 of the 26 three-lane sections the traffic flow would be between 50,000 and 79,000 vehicles a day in the 15th year of operation, whereas the flow within a very short time was between 81,400 and 129,000. In the 1990s, a series of studies in the US, including those by the likes of Kain and Pickrell brought into question the forecasts of transit ridership and costs of public transit investments; and there is little evidence that they have improved subsequently. More recently a series of studies by Flyvbjerg and associated co-authors looking across a range of surface transport forecasts has shown considerable inaccuracies extending to most western economies, and provide confirmation of the poor performance of forecasting models. There emerges a particular tendency for over-prediction of capacity utilization and under-prediction of the outcome costs of investments – to take a few example, for ten rail projects from a variety of countries, the passenger forecasts overestimated traffic by 106 per cent, whereas for road projects there is a noted tendency for the forecasts to be wrong by about 20 per cent but the errors were spread equally around the ultimate flows. In terms of costs, an examination of 58 rail projects indicates overruns averaging nearly 45 per cent, and for 167 road investments, overruns of 20.4 per cent.

If these forecast were a major factor in both the decisions to make the investments and in their design, then this would imply a serious miss use of resources across modes, and probably between transport and other activities in the economies concerned. The problem may in fact, well be considerably worse than the quantifications by Pickrell, Flyvbjerg, and others in that there is no way of knowing if forecasts of investments that were rejected were biased and instrumental in the rejects. If so then resources that would have earned a social return in those projects would have been transferred to some other use where the net benefits are less. We only have data on the investments that actually materialized.

The Causes of the Problem

There are three broad and entwined reasons why transport forecasts are generally poor: state of knowledge about forecasting, carrying out the forecast, and using the forecast. The importance of each is not constant, but can vary with circumstances over time. The most transparent reason why forecasts are inaccurate is that the techniques used are deficient in some way. No one would expect perfection in forecast, there are too many uncertainties bit one would expect some improvement in accuracy over time. There seems to be no lack of discussion of methodologies with numerous national and international conferences that are continually being held where, often admittedly exactly the same, experts gather to discuss their work and findings, and there are a plethora of academic journals and monograph series where, admittedly largely the same people, peer referee and publish their work. Communication on recent findings and best practices, does not, therefore, seem to be lacking, nor do the mathematical skills required to do the technical work seem in short supply.

The advent of computers and the development of specialized software have made data manipulation easier and the testing of hypotheses less cumbersome. Technology, therefore, seems to pose few constraints. Data is always a problem, but even here there have been developments. We have better survey design techniques, and there are more ways to gather transport related data ranging from on-line surveys to the use of global positioning systems for tracking movement. It is true that the move towards lighter regulation and privatization of transport facilities has often reduced the public sources of data, but this is a relatively minor problem compared with all the advances just noted.

The problems in many cases would seem to lie more in the way forecasting is actually done and the results used, two problems which often go hand in hand. A problem initially highlighted Kain is that forecasts are not politically neutral. Many decisions regarding transport investments and policy are not made in the public interest, but rather at least to some degree serve the ends of those who are making them. Those making decisions have their own agenda and can manipulate the processes in which forecasts are made and used. Basically, the forecasting process and output can be captured, to adopt the jargon of regulatory economics by those who commission the forecasts and then make use of them.

These may be politicians who wish to win re-election and thus positively assess the short-term gains of supporting high-use/low cost investments provided by some forecasts against other less "optimistic" projections, or they may be bureaucrats concerned with increasing public sector activities. Some decision-makers simply have a "monument complex" and accept forecasts that favour large, capital-intensive options. One problem is the possibility of game playing, with the authorities consciously keeping costs estimates low at the outset to get a policy accepted but knowing that subsequent and inevitable modifications will lead to cost escalations. The system is, in strict terms, corrupt and even if economists and engineers are neutral in their work, this has little influence on the way their output is interpreted and used. Thus to try close the gap between forecast and actual outcomes is more a matter of institutions than science.

The Potential Treatments

There are a number of possible options for handling the poor forecasting that accompanies many transport appraisals. The options may not be viable or useful in all contexts, but an informal SWOT (strengths, weaknesses, opportunities, threats) style analysis is helpful in isolating which alternatives have merit and in what situations, as well as offering some insights into the practical problems of adopting them.

Do-nothing

In some cases more accurate forecasts may also not be required even if the existing techniques have shortcomings. For example, the poor performance of national car ownerships for the UK in the 1970s and 1980s, which were an integral part of the traffic forecasting of the time, really had little impact on decision-making. Road capacity was clearly inadequate and whether traffic was to grow at 2 per cent a year or 4 per cent made little difference because of budget constraints. More important was where the demands were likely to be strongest, but even here the trends were manifestly clear, and accurate, detailed projections really would not have helped in prioritization. As another example, when road congestion charging was initiated in London in 2003, a round sum of £5 a day was selected as the charge. The need for exact demand elasticity was not seen, but the need to implement the charge expeditiously was, and the price on access to London at peak times produced a significant improvement in the city's traffic conditions. The potential threat of the do nothing situation is that it will be extended to transport projects where more refined and accurate forecasts are needed for good policy making.

Improving the Forecast Methodologies

The theoretical methodology of transport modelling has gradually evolved from the engineering-based four stages – generation, distribution, mode split, assignment - to a greater appreciation of the importance of human behaviour in travel patterns. New concepts include agent based analysis, behavioural, and economics. However, translating these concepts into practical tools for forecasting is not easy, and this may explain why the four-stage process is still widely used. There are challenges in the estimation of parameters, collection of apposite data, prediction of the future paths of independent variables and so on.

The approach of the modellers themselves is not always helpful, and while much attention in the academic literature is paid to how well the newer models can explain prior travel patterns, there are few articles that provide actual forecasts that can subsequently be evaluated (perhaps because academic tenure and promotion have little to do with the practical needs of the transport planning community). It is often more difficult to foresee some trend break, or some new influential trend in factors affecting transport demand, than it is to estimate parameters from historic data that implicitly assume on-going trends continue. One new technique does show great promise. Nobel Prize winning economist Dan McFadden deployed a disaggregate discrete mode split model to examine the construction of the Bay Rapid Transit System (BART) in the San Francisco area. While the conventional aggregate gravity model forecast a 15 per cent modal share for BART McFadden's disaggregate forecast was 6.3 per cent and the actuality was 6.2 per cent. Unfortunately, despite this impressive performance, BART has never adopted disaggregate modelling as a policy tool.

Even when new techniques prove themselves, there are major barriers to their adoption. In the US, for example, few states have senior economists and are dominated by civil engineers who are more inclined to treat "human factors" as somewhat secondary when looking at traffic flows. The situation is similar in other countries. In many cases there are also significant costs involved in changing forecasting models. The forecasts are often an integral part of a wider planning and design process embracing interactive suites of programs. The problem is potentially compounded by the capture of much of the system by large consultants interested in selling their four-stage modelling software.

Correction Factors

If the forecasting biases are symmetric and have approximately the same order of magnitude across studies, then it may be able to apply an adjustment factor to correct the errors. For example, if the forecasts always underestimate costs by 20 per cent then inflating the forecasts by 25 per cent will give better forecasts. The UK Department for Transport has explored this option along with other departments. Some sort of benchmark is adopted against which forecasts may be assessed; serious deviations from these benchmarks require specific justification. This approach is relatively easy for policy makers to understand and has found some favour in a number of government departments.

The major challenge is in defining what the benchmarks should be and drawing the boundaries for when a particular set of forecasts are called in for detailed scrutiny. Who makes the decisions that the forecasts are over-optimistic and the appropriate correction to apply? There is the additional danger of game playing, with those producing the initial forecasts of, say, costs, making even lower forecasts to which the correction factor will apply. There is also the question of how to handle forecasts of novel policies or technologies where benchmarks really do not exist. Overall, there is always the threat that any corrective process to correct is itself captured by those with a strong interest in a particular outcome.

Bringing more Views into Forecasting

Traditional engineering-based techniques tend to produce single forecasts that give limited ideas of the ranges of uncertainties involved or the potential sources of these problems. Sensitivity analyses are sometimes offered, but they relate to risk, where there are pretty well-known error distributions associated with parameters, rather than uncertainty where such distributions are highly opaque. Added to this, the forecasting methodologies are often almost incomprehensible to laymen because of their technical nature.

There are two possibilities for greater involvement in forecasting: more use of expert opinion and more public participation. In the former context, a number of private transport entities have embraced the idea of supplementing "technical" forecasts with adjustments following consultations with experts. Boeing Commercial Airplane started doing this with it traffic forecasts in the

1990s as did British Airways. This may not satisfy the statistical purist, but it does offer the opportunity for introducing insights into possible trend breaks or new trends in traffic patterns.

Regarding the second option, making the forecasting methodologies more transparent and the assumptions more open to debate has been a long-standing objective. One major challenge to this objective has been the advent of standardized appraisal methodologies in the UK, starting with common discount rates across sectors in the 1960s and moving in the specific transport context to COBA. The on-going problem is to make operational something upon which there is considerable agreement; basically to demystify forecasting and open discussion of the possible scenarios to include in the forecasting process.

Engaging the Privates Sector More

There is considerable evidence that private sector forecasts are more accurate than in the public sector, suggesting the value of shifting transport infrastructure decisions to the private sector. In one study of 183 transport projects by Flyvbjerg, Holm and Buhl, the private sector ones ultimately experienced cost escalations of 34 per cent and public sector ones of 110 per cent. More generically, and as an official response to over optimistic forecasts, the UK Treasury has offered guidance on how public sector agencies may cope with them following the publication of the Mott MacDonald Report.

The difference in the quality of public and private forecasts can be understood in the context of the respective principal-agent situations. Since the spread of stakeholders in the private sector is much less than the public sector, where any adverse consequences of inaccurate cost or revenue forecasts fall on all tax payers, decision-makers in the private sector tend to be more conscious of the risks of making mistakes at the project level. The agents are more aware of the interests of their principals in private companies, although admittedly not always entirely responsive. There are fewer ways, for example, to disguise a mistake by bundling it with a large number of other projects, and the costs of poor forecasting are borne by a narrower range of stockholders who tend to be more focused. By shifting more investment decisions to the private sector, there is a natural incentive for those involved to be more careful in their forecasting. The incentive for either padding predicted usage or understating costs is diminished when there is close accountability. This shift of responsibility can be done through a variety of mechanisms including more purely private investments, private sector concessions, or requiring public agencies to seek private sector support for any of their own projects that fail to meet forecasted outcomes. One weakness of this approach is that the private sector often has more difficulty financing major projects. Another problem is that a private sector provider may exercise monopoly power once the project is built. But there are ways of dealing with this, including various price capping regulations and a plethora of options around the build, operate, and transfer framework.

More Flexible Engineering

Another response to the problem of poor forecasting is to accept the inherent uncertainties in very long-term projects and adopt a more flexible strategy towards infrastructure construction. This may be particularly germane regarding large, durable projects. The normal discounting procedures used in techniques such as cost-benefit analysis have in the past meant that any errors in long-term projections were weighed lightly in decision-making. The increased concern about sustainable development, and in particular the Stern Report's recommendation that very low discount rates should be used in appraisal means that long-term effects play a more significant role in cost-benefit assessments. Poor forecasts effectively have a larger impact on the welfare calculus. Greater flexibility could be introduced by, for example, leaving rights of way along roads for possible capacity expansions or the introduction of alternative modes. This is essentially an engineering design approach that allows for phased construction with the possibility of modifications to meet emerging circumstances as time passes. It has the strength that it can minimize any adverse implications of "corruption" in the initial forecast and allow for unforeseen events and to adopt new technologies or concepts as they emerge. Against this, there are possible costs of adopting designs that are not in the short term the most productive. Another danger is that forecasts required for subsequent decisions regarding modifications will themselves be captured by vested interests.

TRANSPORT'S CONTRIBUTION TO ECONOMIC PROSPERITY

BETTER USE OF TRANSPORT ASSETS

Underused infrastructure and other transport assets are an inefficient use of resources and a drain on economic performance. Smarter use of existing transport assets can improve efficiencies and reduce or postpone the need for additional investment. We can get better use of existing assets by managing transport demand across the network and by improving the management of traffic flows. This can involve: the use of information and communication (ICT) to inform travel decisions; incentives to travel on different modes or at different times; better links between transport modes; the use of technology to improve freeway flows; and the use of road hierarchy plans such as SmartRoads to improve flows on prioritised transport modes or routes.

IMPROVE ACCESS TO WORK AND EDUCATION

The transport system plays an important role in the development of human capital, itself a major driver of prosperity. Transport connects people to education, training and skills development. An efficient transport system also increases the number of potential workers within a reasonable travel time of each workplace. This deeper pool of labour improves the potential to match a person's skills and career aspirations with an employer's workforce needs. It also improves employees' ability to find more complex and better paid roles as their skills develop.

Support Business Clustering

Businesses usually locate near one another in central business districts, retail centres, industrial and freight precincts, and research and education facilities. Co-location allows them to share public infrastructure, improve their supplier base, gain efficiencies through specialisation, spur creativity and innovation through knowledge sharing and horizontal linkages among firms, and minimise freight movements. These factors boost productivity and prosperity.

The clustering of business activity can also create costs such as congestion. A transport system that minimises the costs of business clustering will support the productivity that comes from proximity. This may involve ensuring efficient transportation for workers travelling to the central city, providing reliable and efficient transport to commercially attractive land, and ensuring good transport access to knowledge precincts such as universities and research clusters.

Provide Value for Money Infrastructure and Services

Prudent and reliable management of the state's finances is essential to Victoria's economic prosperity. Investments should provide value for money and should only be made where lower cost alternatives are not available. Investments should also be coordinated across levels of government and provide certainty to the private sector.

Transport infrastructure investments are amongst the most expensive that governments make. The transport portfolio will ensure these investments are well-planned, rigorously appraised, focussed on objectives, scheduled and prioritised appropriately, and designed to support efficient and productive land use. Private funding options will be pursued where possible.

Improve Business Access to Markets

Transport improvements allow businesses to more efficiently move their goods and services and to access a wider range of suppliers and customers. Victoria is a major transport and logistics centre for south-east Australia. Reducing bottlenecks in supply chains and improving access to key economic nodes such as ports, airports and freight distribution centres will improve market access. Business travellers and customers place a high value on time, and a transport network that provides efficient connections between businesses and their customers will improve competition and prosperity.

Keep Transport Costs Down

The freight transport and logistics sector contributes about 10 per cent to the Victoria's economic output and transport costs form a component of the final cost of all goods and services. Transport costs form 16 per cent of household expenditure. The productivity of businesses and the prosperity of individuals and families can be improved by minimising transport costs. These costs can be inflated by congestion, inefficient regulation or poor reliability of the transport system.

Improvements to the transport network and effective traffic management measures can lessen delays and costs incurred by freight carriers and business travellers, and provide greater reliability and on-time running for service deliverers. Well-targeted and efficient regulation will also minimise costs, for example by removing cross-border regulatory anomalies. Where there is market failure, measures to stimulate greater competition can also reduce costs.

TRANSPORT'S CONTRIBUTION TO SOCIAL AND ECONOMIC INCLUSION

RESPOND TO USER EXPECTATIONS

Transport users want a reliable transport system that is safe, efficient, comfortable, and accessible. It is important to make sure that transport can be used by as many people as possible, and that the whole trip has seamless connections and is people-focussed.

Make Transport More Widely Available

Achieving the best transport outcomes requires transport infrastructure and services to work together as part of a single integrated transport system that reaches as many people and places as possible. This includes considering frequency of services. Victorians need a wide range of transport choices that are responsive to different abilities, preferences and purposes.

Improve Transport Affordability

The high cost of motor vehicle transport, in particular, impacts on the cost of living for many families. The cost of travel can be a significant barrier to accessing social and economic activities.

This may include direct costs, such as fares and fuel, or associated costs like licence renewal fees and registration. The provision of transport concessions – for car registrations or public transport fares – are one way transport can be made more affordable.

BUILD CAPACITY

Engage and Collaborate in Planning and Delivery

Providing opportunities for individuals, communities and organisations to input into transport planning will assist in delivering better transport solutions, and a safer and more reliable transport system, by engaging with the people who use or are affected by the decisions of transport agencies. Contemporary challenges such as population growth and increasing traffic congestion require a more integrated response. Collaboration with all levels of government, the transport industry and transport agencies is key to effectively responding to complex challenges.

Support others to Take Action on Transport Challenges

A key challenge is to harness the capacity of communities, businesses and local councils to develop transport solutions and mobilise local transport assets such as community buses. In some circumstances the government can play a facilitation role to support private and not-for-profit players to introduce transport solutions, particularly those focused on responding to local transport demand.

Create a Positive Legacy

There is considerable scope for transport projects to contribute to broader outcomes above and beyond transport outcomes. For example, infrastructure development and maintenance can contribute to local economies and employment, and through engagement and partnerships, contribute to local knowledge and leadership. By providing more opportunities for people, the transport system is able to create a positive legacy; that is, ensuring projects have a positive impact that will remain in the community after the transport infrastructure or service has been delivered.

TRANSPORT'S CONTRIBUTION TO RESOURCE EFFICIENCY AND ENVIRONMENTAL SUSTAINABILITY

Reduce Distances Travelled to Access People, Places and Goods

Land use and transport are interdependent. The design of the transport system and land use decisions affect the overall demand for travel, the length of trips and the modes that people choose. Reducing how far people and goods need to travel (without affecting people's ability to access the things they need) reduces emissions and congestion and improves the efficiency of the transport system.

Make Transport Activity More resource-efficient and reduce its Environmental Impacts

Current technology can reduce the fuel used by road vehicles by at least

30 per cent. This has the added benefit of reducing their greenhouse intensity. Over time, an increasing number of cars and trucks are expected to be powered by alternative forms of energy. Supporting more efficient vehicle and fuel technologies will be vital for minimising our impact on the environment. The greenhouse performance of our trains, trams, buses and taxis can be improved by using more renewable sources of energy and adopting more energy-efficient technologies and practices.

Air pollution from motor vehicles continues to pose a significant health burden. We need to consider what more should be done to reduce this burden and to manage the noise impacts of transport activity.

Use Environmentally Sustainable Transport More

About 60 per cent of passenger trips are less than 5 kilometres in distance. Supporting mode shift from single occupant cars to lower-emissions options such as public transport, cycling or walking will increase the resource efficiency and environmental sustainability of travel without reducing access to economic and social opportunities. Increased use of rail to move freight can reduce emissions and road congestion, and help to conserve oil and gas resources.

Make Transport Infrastructure More resource-efficient and reduce its Environmental Impacts

The construction and management of the transport system requires water, fuels, minerals and land. Making efficient use of our resources today will help to ensure future generations will also have the opportunity to enjoy comparable levels of prosperity. Minimising harm and improving environmental outcomes from construction and management of the transport system is essential for protecting biodiversity and natural eco-systems including flora, fauna and habitats.

Make Transport Resilient to Climatic Extremes

Climatic conditions are changing. The transport system is vulnerable to extended periods of extreme heat, more frequent and more intense storms, flooding and the impacts of drought on soils. The design and construction of infrastructure today will determine the capacity of the network to withstand future climatic conditions. Using historical data as the basis for future designs and plans presents a significant risk to infrastructure owners and operators in a changing climate. Collecting new information and supporting the transport sector to take new climatic factors into account will be critical to developing a resilient and reliable transport system.

3

Transportation Planning

Transportation planning, or transport planning, is involved with the evaluation, assessment, design and siting oftransport facilities (generally streets, highways, bike lanes and public transport lines).

CONSENSUS REGARDING THE PRIMARY TRANSPORTATION DOCUMENT

There is consensus among the transportation agency stakeholders that Houston Galveston Area Council's Regional Transportation Plan (RTP) is the dominant transportation work driving the majority of transportation decision making for the Greater Houston Metropolitan Area. The RTP includes planned projects for TxDOT, 8 proximate counties, municipal governments, and other entities. (In various years, the plan may be named the Metropolitan Transportation Plan, but the document is the same.) Other documents, which may emanate from or contribute to the RTP are the City of Houston Major Thoroughfare Plan, METRO's Solutions (and prior to November 2003, the Regional Bus Plan, plus the Main Street Light Rail), the Harris County Toll Road Plan and other projects submitted independently by public entities. At least 18 planning and project reports may influence transportation in the region. The most available and well-known of these documents are described in Table. The primary plans, with the greatest influence on the RTP are discussed in Section 3 and those with a secondary influence are described in Appendix A.

Transportation planning is hierarchal with the RTP as the umbrella long range document. Projects receiving federal funds must be included in the RTP. Local projects not using federal dollars are not required to be in the RTP, but are included.

As part of the background transportation system. All known transportation projects are assessed for the projected number of users (called travel demand), approximate determination of capital costs, and impact on air quality. More detailed planning and engineering work occur thereafter with the complexity and level of detail increasing as the planned projects move toward completion.

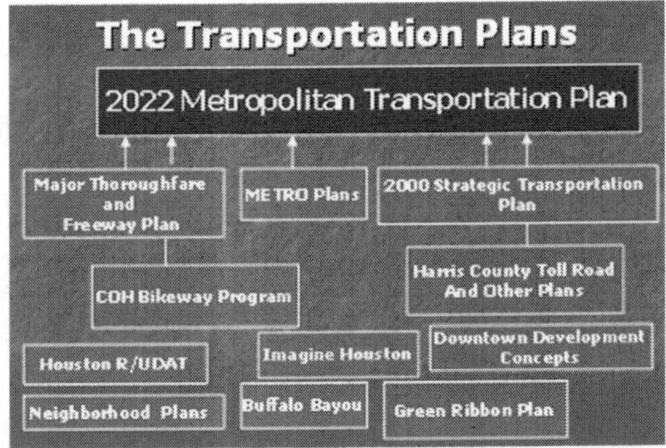

Fig. Regional Transportation Planning Framework

Federal guidelines exert great influence over the local transportation planning process. Regulations and guidelines come from the Transportation Equity Act for the 21st Century (TEA 21, currently undergoing reauthorization by Congress) and the National Environmental Policy Act (NEPA). Any transportation project receiving federal money must comply with the procedures and sequencing described in these Acts. Requirements for public involvement are one of many mandates included in the federal regulations. In addition to the formal federal guidelines, the federal elected officials also may influence transportation projects and funding, outside of the formal planning process.

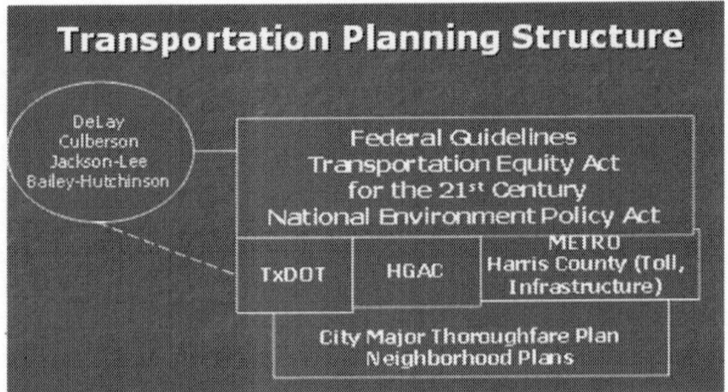

Fig. Federal Influence in Transportation Planning

HOUSTON GALVESTON AREA COUNCIL'S REGIONAL (OR METROPOLITAN) TRANSPORTATION PLAN

- *The Plan*: Houston Galveston Area Council is responsible for development of the region's long range plan and for coordination of other agency plans, including the plans for the City of Houston. Over the last years, it has been called the Regional Transportation Plan and Metropolitan Transportation Plan, but the basic parameters and components of the documents are the same irrespective of the title. The plan reflects a compilation of projects from each agency in the region, state government, and other stakeholders, as well as a systematic view of the level of "travel demand." Travel demand is the volume of anticipated trips based on the projections of people living and working in the region. The formally adopted plan is the 2022 Metropolitan Transportation Plan (MTP). The next update titled, 2025 Regional Transportation Plan, is in draft form and public comment has been solicited. Ratification is scheduled for summer 2004.

The Regional Transportation Plan includes new projects, improvements to existing projects, maintenance programs, and non-transportation enhancements that may improve aesthetics or overall operation. The inclusions may be transit, vanpooling or carpooling, and other strategies to influence when people travel (roadway demand) and to make the existing system operate better through techniques such as signalization improvements.

- *The Players*: The HGAC plan is approved by the 24-member Transportation Policy Council (TPC) comprised of elected county and city officials, TxDOT and county engineers, METRO and selected other representatives. The TPC is advised by the Technical Advisory Committee (TAC), a 31-member technical group of planners, engineers, port officials and stakeholder constituent groups.
- *Citizen Involvement as Described in the Document*: The public involvement process follows the typical transportation agency format of called pubic meetings, solicitation of written comments, newsletters and web-based information dissemination. The public may also address the TAC or TPC during an open comment period at the beginning of each scheduled meeting. Historically public involvement has been low and generally focused on citizens discontented with specific projects, rather than the plan as a whole. HGAC proceeds according to a published public involvement plan approved by the TAC and TPC.
- *The Process*: HGAC requests the most recent set of plans from the transportation entities within the region, which forms a portion of the RTP's foundation. Principal inputs include the City's major thoroughfare plan and Capital Improvement Program (CIP), bicycle master plans, METRO's plans and anticipated projects, TxDOT's short and long range Unified Transportation Program, the state implementation plan for air quality and the 3-year Transportation Improvement Program. Also, individual projects are identified and submitted by a sponsoring agency (*e.g.*, METRO, TxDOT or a municipality). A project description, which documents the need, the number of projected users, the cost and other relevant information accompanies the submittal. Once submitted for inclusion in the RTP,

the project is assessed by the HGAC staff based on predetermined criteria and may be accepted or not. If accepted and included, the sponsoring agency will continue with more detailed studies and two stages of engineering, preliminary and detailed.

Projects may also be identified as the result of a needs assessment by HGAC, which identifies system deficiencies or otherwise notes projects that would improve transportation system flow or operations. Projects arising by this method would need to be accepted by a municipality or governmental agency in order to proceed because HGAC plans, but has no implementing authority.

The total proposed regional system is tempered by the funds available for capital/construction costs and maintenance costs. The system is not as large as the demand might indicate if more funds were available for construction or operation. Simultaneous to the entire development process, citizen involvement should be on-going.

The inputs are combined in the RTP document to form a comprehensive overview of area plans and projects. The document includes vision and goals statements formed from comments at public meetings and through opinion surveys. The plan is assessed against a financial background of affordability, and computer models of travel times, routes and speeds are processed. The computer outputs show the areas of congestion both with and without the plan implementation. Conformity or nonconformity according to the air quality requirements is determined. The plan is presented to first the Technical Advisory Committee and then to the Transportation Policy Council for approval.

The first three years of the Regional Transportation Plan are identified as the Transportation Improvement Program and forms the basis for the implementation and construction. During engineering and construction stages, the HGAC TPC and TAC are provided periodic updates of major undertakings, such as freeway widening projects, new roadway construction, and transit plans. Some smaller roadway projects may be presented to the TPC and TAC, as well, particularly if more funding is desired or there is controversy associated with the project.

The Results of the Plan Produced

The plan is very effective at shaping implementation of projects in the region. Any project receiving federal funds must be included in the long range plan.

Thus, agencies are careful to ensure that desired projects are incorporated into the 2022 (or upcoming 2025) plan. The first three years of projects included in the long range plan are culled to form the Transportation Improvement Plan, which is the basis for the agencies to request imminent funding from the federal transportation funding sources.

SYNOPSES OF PRIMARY CONTRIBUTING PLANS TO THE HGAC REGIONAL TRANSPORTATION PLAN

Although there are many plans that contribute in some way to the region's transportation systems, those prepared by TxDOT, the City of Houston, METRO and Harris County form the basis for most projects and activity. The process of developing the individual and agency components is complex, involving many governmental layers and other organizational stakeholders. Neighborhoods, individual developers, and other interest groups petition the transportation agencies, either advocating or opposing specific projects. The participation in the general planning process has tended, in the past, to attract few interested citizens. There is an indication that may be changing as a coalition advocating transportation choice and improved planning has recently been formed. The following text describes each primary plan development process, including the players, citizen involvement and results of the plans.

City of Houston Major Thoroughfare and Freeway Plan

- *The Plan*: The City of Houston's Code of Ordinances authorizes the Planning Commission to prepare and adopt various types of master plans, which involve the physical development of the city. The Major Thoroughfare and Freeway Plan is such a plan. First developed in 1942, the City updates the Major Thoroughfare and Freeway Plan (MTFP), annually. The plan considers comments provided by developers and neighborhoods about congestion, mobility and future development plans. Sections of roadways, which are in need of expansion, lengthening or widening are identified. The plan serves as notice to the public of future road development and the public's requirements for developing land adjacent to the identified roads. The outcome is a City Council Resolution with a map, adopting those amendments. Distribution is via the web and includes the map, the policy report, and the amendments.
- *The Players*: The City Planning and Development Department is responsible for the Major Thoroughfare and Freeway Plan. The plan is reviewed and accepted by the Planning and Zoning Commission and approved by City Council. Meetings about amendments to the major thoroughfare plan are held with agency stakeholders including the City's Public Works and Engineering Department, Harris County engineers, Montgomery County engineers, METRO, Harris County Toll Road Authority, and the Grand Parkway Association. Also, the Planning and Development staff emailed applications to TxDOT for review.
- *Role of Citizens*: A mailing is forwarded to property owners along current and proposed alignments of the map, which contains the draft case report with the specific amendment pertaining to their

property. Thereafter, a public hearing is held. Amendments to the thoroughfare plan can be proposed by the private sector or by staff in conjunction with other government agencies and are posted 7-10 days before the public hearing.

- *The Process*: Applications are received during a 4-week period, roughly February to March, with the Planning Commission public hearing held in July and approval by City Council in September. Recommendations are made by the Planning and Development Department; the Commission accepts or requests changes.
- *The Results of the Plan Produced*: The Major Thoroughfare and Freeway Plan forms the foundation of the City's 5-year Capital Improvement Program and is an important component of the HGAC 2022 Metropolitan Transportation Plan.

City of Houston Capital Improvement Program (CIP)

- *The Plan*: The Capital Improvement Program is a plan for physical improvements to public facilities and infrastructure. It is principally a listing of projects and budgets with explanation of the street classification system. In November 1983, the City Council established the fire-year capital improvement plan process. By resolution, it became City policy to engage in a continuous five-year capital improvement planning process that includes annual reviews, revisions, and adoption of a five-year CIP. The CIP is revised annually to include new projects, reflect changes in priorities, and extend the plan an additional year. The first year of the plan is the current capital budget and is revised throughout the year as needed or when changes are made to existing approved capital projects. The financing of capital projects covers land purchase, construction of new facilities, initial equipment purchases to furnish new facilities, and all related planning, engineering, and architectural design activities.
- *The Players*: The CIP is developed by the Department of Public Works and Engineering. The Mayor, City Council members, and other departments take a visible role in shaping the construction schedules and the projects that are included in the CIP.
- *Role of Citizens as Described in the Document*: While not noted in the document, most Council Members hold meetings in their districts for citizen comment about the CIP. Also, citizens may provide considerable input to their respective member of council in an informal way through comments provided independent of the formal meetings.
- *The Process*: The process, beginning in February, follows an abbreviated schedule with elements similar to those of previous

years, City departments begin at that time to review projects in the current CIP to determine whether rescheduling of projects and adjustments in funding will be required in the next CIP. In March, public meetings are conducted in all Council districts. At these meetings, coordinated by the Planning and Development Department, comments are received regarding capital project plans and respective City departments. In April, a number of departmental CIP submissions are received, and in May, the remaining submissions are received, reviewed, and evaluated by the Finance and Administration, Planning and Development, and Public Works and Engineering Departments. CIP reviews and discussions between departments and the CIP Coordinating Committee are held in June. The schedule for projects is set based on funding availability, need, and comments received from Council.

- *The Results of the Plan Produced*: Upon completion of the CIP submission review sessions, a proposed CIP for the next five years is prepared for presentation to City Council for review and adoption. The projects in the CIP form the basis of the roadways constructed in the City and the CIP is incorporated into the 2022 MTP and some of the projects are included in the 3-year Transportation Improvement Program.

METROPOLITAN TRANSIT AUTHORITY PLANS

The Plan: METRO develops long range plans, which include projections for bus routes, identifies corridors for high capacity bus and carpool lanes and rail. The 2022 HGAC metropolitan plan included buses on high occupancy vehicle lanes (HOVs) for all freeway corridors and the 7.5-mile light rail line. In 2003, METRO presented a plan for light rail expansion, more bus routes, more express buses and a continuation for at least five years of providing 25 per cent of its sales tax to the City of Houston, Harris County and municipalities within METRO's boundary. METRO area voters ratified this plan in November 2003. Most METRO capital and construction projects utilize federal money. The Players: Metro staff and consultants prepare the technical needs assessments and alternatives that respond to anticipated transit ridership levels. HGAC, TxDOT and Harris County participate in periodic review sessions at key points in the planning process. However, the coordination is not consistent and occurs as the need for coordination becomes apparent.

The Process: The toolbox of transit options (bus, express bus, bus in HOV or rail) are applied based on computer generated projections of transit ridership. The possible transit option is tempered by the funds available for capital/construction costs and the costs to operate the future system. The system is not as large as the demand might indicate if more funds were available for construction or operation.

Citizen Involvement As Described In the Document

Involvement from the public occurs via public meetings regarding bus service changes, meetings and hearings for the corridor studies, web page linkages, focus groups and speaker's bureaus. Specific outreach for the METRO Solutions Plan, subsequently approved by voters included three sets of public meetings. The first set occurred in January and February of 2003. Thirteen meetings were held throughout the region and comment cards were available on the buses. METRO distributed over 100,000 meeting notices for these meetings. The second set of 15 meetings occurred in May 2003. The transit agency mailed 400,000 brochures describing the options, which had a detachable mail-back comment card; 16,000 cards were returned with comments. Meeting notices were again displayed on the buses. METRO held the final set of 19 meetings in September and October 2003. Another 400,000 brochures were mailed, as was a small reminder card of the referendum components.

Prior to the 2003 meetings about METRO Solutions, from about fall 2001 to December 2002, METRO held numerous corridor meetings, focusing on the North, Uptown/West Loop and Southeast areas slated to be the next rail corridors. In particular the meetings for the North Corridor were widely commended as a very good example of citizen involvement. METRO and TxDOT worked together in this corridor. There were also strong community and civic organizations active in distributing communications and working with the public agencies. The consultant hired by METRO is credited by many persons for setting the tone of the public involvement for this corridor.

Results of the Plans Produced

The plan components are included in the HGAC Regional Transportation Plan. Individual corridors and projects undergo more detailed planning and engineering and environmental assessment leading to implementation.

TEXAS DEPARTMENT OF TRANSPORTATION

The Plans

Several processes and resulting documents frame TxDOT's transportation proposals including formal avenues such as major investment and corridor studies and informal means, such as a suggestion by an elected official. In the formal set of documents, the most comprehensive are the short and long range Unified Transportation Programs, which serve as TxDOT's key submittal for the RTP. Specific corridor projects undergo Major Corridor Feasibility Studies and detailed engineering. An example of the informal path is reflected in changes made to the Katy Freeway expansion. Original designs included a 4-lane, bi-directional High Occupancy Vehicle (HOV) facility in the center of the reconstructed freeway. Largely through an initiative spearheaded by the

County Judge and a US representative, the middle lanes were transitioned to a toll road. The roughly $250 million dollars advanced to TxDOT by the Toll Road Authority speeded the construction schedule for the project by about 2 years.

The Players

The key players for TxDOT plans are that agency's officials and employees, HGAC, and elected officials. For some corridors and projects, other agency stakeholders, such as METRO may be engaged.

The Processes

Roadways emanating from TxDOT are almost always a reflection of technical engineering methodologies. The major corridor investment studies include public involvement per federal guidelines.

Citizen Involvement

The public involvement process follows the typical transportation agency format of called pubic meetings, solicitation of written comments, newsletters and web-based information dissemination. The number of citizens participating in the meetings may vary from those with low interest (a recent meeting in the 225 highway corridor drew 5 attendees) to those meetings that attract several hundred participants.

The level of public involvement is higher when there is dissent or discord around the proposed projects.

Results of the Plans Produced

The plan components are included in the HGAC Regional Transportation Plan. Individual corridors and projects undergo more detailed planning and engineering and environmental assessment leading to implementation

HARRIS COUNTY TOLL ROAD AND OTHER ROADWAY PLANS

The Plans

Harris County Commissioners have responsibility for development of roadways and infrastructure within their precincts. The Department of Public Infrastructure conducts the engineering and implementation phases. Harris County responds to the solicitation of call for projects by HGAC that goes to local municipalities and transportation agencies. Harris County also provided projects termed Toll Road Pooled projects, which accounted for the toll road portion.

Harris County has a Capital Improvement process, but their CIP is not forwarded to HGAC in its entirety; rather, Harris County submits its desired projects via HGAC's solicitation process.

The Players and the Processes

Harris County technical engineering staff assembles the project lists and descriptions, in consultation with the Harris County Commissioners. Project funding for each project is through the individual precincts. The citizen contributions are unclear. Results of the Plan Produced: The lists forwarded to HGAC are included in the RTP. Harris County also constructs many roadways that do not have a federal financial component.

SYNTHESIS OF INTERVIEWS

The study team conducted interviews with 11 stakeholders representing each of the principal agencies involved in transportation planning and other individuals thought to represent active citizens' perspectives.

The interviewees concurred that HGAC is the lead institution in the region for transportation planning and most indicated that the Regional Transportation Plan is largely followed. Projects have been deleted from the plan in years past, but essentially the RTP provides the basis for development. Most contend that each agency conducts its individual planning efforts and HGAC compiles the components. There is not a coordinated structure facilitating trade-offs of one agency's projects vis-à-vis projects by another agency. The process is focused on projects, rather than on planning.

- The City of Houston should become more active in the transportation planning arena.
- Coordination between agencies occurs, but not necessarily in a coordinated, consistent, structured or formal manner.
- Developers are extremely influential in the planning process and are knowledgeable about the process.
- There is not adequate discussion about the role of the various planning documents in the regional planning process.
- The current organizational structures inhibit new ideas entering the process, as well as changes to existing plans.

Suggestions relative to improving the transportation planning process are as follows:

- The City should have a transportation czar responsible for coordination and advocating for the city.
- Planning should be instituted in the City, which considers the transportation and land use interface.
- There should be a coalition between the urban and suburban core.
- Citizens should be consistent in their involvement.
- Agencies should form a coalition, with technical loaned positions from the County, City, METRO and HGAC

The consensus is that the planning process works relatively well, given the number of stakeholders and agencies, and breadth of the transportation system and components. Still, improvements could be made in regular, systematic coordination among agencies and stakeholders and in increasing the comprehensive nature of transportation planning.

AN EXAMPLE OF RECENT CITIZEN AND PUBLIC AGENCY COORDINATION

The closing of Spur 527 on the Southwest Freeway provides a recent example of citizen involvement from the perspective of the proximate residents and businesses and from the perspective of TxDOT, the sponsoring agency. Nearby residents expressed much discontent about the construction plan, which called for closing the primary exit to downtown, Spur 527. This closing worried the residents as the engineering analysis showed significant traffic flowing on local streets due to the ramp closure. A Texas Southern University student, Jermaine Hannon, investigated the history of this project with consideration of the public's participation, which began in the early 1990s. At that time, TxDOT initiated plans to reconstruct the freeway section from Kirby Drive to its intersection with IH 45.

Early plans proposed an elevated freeway section, which the nearby neighborhoods opposed. During the early to mid 1990s, numerous meetings were held with the neighbors to work out details of the construction. TxDOT agreed to depress the freeway section and redesigned the bridges to suspension, which improves the aesthetic appearance of the facility in that section.

The residents seemed satisfied with the accommodations made by TxDOT. The official transcript shows only one comment was made about closing Spur 527. In 2003 as the time to begin construction approached, the issue of closing the Spur became a major problem. Hannon's work indicated the following findings.

- So much time elapsed between the successful meetings and construction that many of the people involved in the citizen involvement process no longer lived in the neighborhood; those that remained had turned their attention to other matters, no longer interested in the project, once the issue of elevation had been addressed.
- New people moved into the neighborhood, who had no knowledge of the prior set of meetings.
- The elected officials involved in the first set of meetings had changed, in large measure due to the City of Houston's 6-year term limits.
- TxDOT had followed procedures for public involvement and proceeded with the project in good conscious, thinking they had worked with the community relative to the issue of the roadway's elevation.
- TxDOT had not maintained contact with the community during the detailed engineering and pre-construction phases, roughly taking place between the late 1990's and 2001/2002.

SUMMARY FINDINGS AND RECOMMENDATIONS

Transportation planning is on-going, undergoes changing methodologies, and is conducted by numerous agencies and civic groups. The study team collected and reviewed more than 18 transportation documents and reports. Summary findings are described below:

- The format and contents of the plans vary widely from those that simply provide goal statements with no specifics suggesting response to the goals to those with detailed project descriptions and accompanying impact statements.
- The process is complex and hierarchal with HGAC, TxDOT, METRO, and Harris County having the dominant roles in determining transportation components and timing of projects.
- The focus of transportation work in the city and region is project-oriented, not necessarily planning-oriented. Many of the transportation elements are maps and project listings, and in that sense, not a "plan."
- In recent years, the City of Houston has not been as active in the transportation planning arena as it should have been.
- Citizen involvement occurs, but is inconsistent across projects and agencies in terms of the level of citizen interest and agency attention and resources allocated to engaging citizens.
  - Some of these documents are not widely circulated and are difficult to obtain for public viewing and comment.
  - Citizen involvement tends to be most apparent when citizens are unhappy with a project.
- The processes for development of many transportation agency documents occur within the various agencies and result from technical evaluation and internal discussions and are focused on achieving technically set mobility objectives.

Without question, transportation is a quality of life issue. Transportation affects numerous socially important areas, which leads to the many stakeholders and documents that influence the region's transportation. For instance, super neighborhoods must consider transportation as they think through how their neighborhood should function. Activity center organizations (*e.g.*, The Texas Medical Center or Uptown Houston) are at liberty to conduct independent studies to address issues unique to their area at any time. It is probably unrealistic to try to coordinate all of the elements that contribute to transportation in Houston. Still, it is important that the plans "fit" well into the comprehensive city and regional framework. Several items could be considered to improve the flow of transportation discussion and decision-making.

- The City should play a larger role in the regional planning and take a lead in brokering coordination of the elements that affect the City of Houston by:

  - Actively working with HGAC's TPC and TAC, and
  - Through independent assessment and determination of its best interests given the transportation plans of others.
- Citizens should participate early in planning stages and prior to project specific decisions times. They should also maintain contact through planning and engineering stages as project changes can occur with greater engineering detail. It is important that citizens recognize the perspective of the agency in trying to gauge the entire spectrum of public opinion, including the silent majority.
- Agencies should spend more attention and resources maintaining contact throughout the planning and engineering processes, even providing updates of planning tasks prior to "answers" being determined. Transportation planners and engineers should focus on the "spirit" of federal citizen involvement requirements, beyond meeting the "letter of the regulations".
- Blueprint Houston can play an important role in establishing the tenor of viewpoints on transportation, expressing a balanced citizen perspective about the direction of transportation in Houston. Blueprint Houston's contribution should be goal and planning focused, not directed to any single project or document.

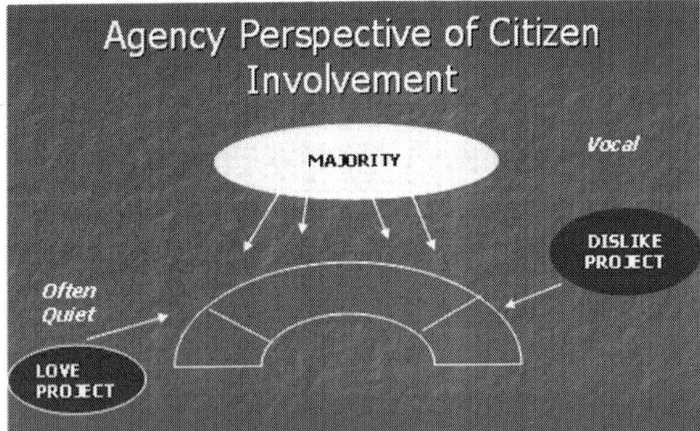

Note: Public agency representatives tend to believe that vocal opposition may not reflect the opinion of the public at-large. In fact, citizens, in private, often tell agency planners that controversial plans are supported by a segment of individuals. The supporters rarely choose to make their opinions known in the public debate, however, often due to feelings of intimidation by project opponents.

THE ROLE OF TRANSPORTATION IN CAMPUS EMERGENCY PLANNING

Campus emergency planning involves many aspects of both the campus community and also the larger surrounding community. Large state universities, like the campuses of the University of California system, may have their own internal response resources, such as a police force. Large private colleges, such as Stanford University, may contract for police, fire and emergency medical services from the local municipal government. Most colleges in the United States are small to medium sized institutions that may have contracted private security services, but usually rely on the local jurisdiction for sworn law enforcement resources, fire response (which may itself be volunteer) and emergency medical services.

Campuses also rely on the local jurisdiction for transportation assets. As described in the monograph The Role of Transportation In Campus Emergency Planning (Edwards and Goodrich, 2009), the campus relies on public transit for evacuation and emergency response assets, and on campus transportation assets for providing immediate emergency shelter, evacuation and Incident Command Post resources. While the campus will generally have its own in-house capability or contracted resources for road maintenance and repair, the campus roads are part of the larger road system of the community, and therefore rely on local and state resources for restoration of service for the roads connecting the campus to the larger community.

In the last five years campuses have begun training students, staff and faculty in the Campus Community Emergency Response Team model (C-CERT), based on the FEMA-sponsored Community Emergency Response Team (CERT) program. Through this program members learn home, workplace and personal preparedness, including workspace and living space mitigation. They learn about building evacuations, fire behavior and immediate suppression (fire extinguishers and mounted fire hoses), management of a hazardous materials accident until the professionals arrive (Recognize, Avoid, Isolate and Notify- RAIN), the care of victims of medical emergencies while awaiting a professional response (Airway, Breathing, Circulation, Shock- ABCs), and search and rescue, such as following earthquake, hurricane and tornado events. (FEMA C-CERT, n.d.a.)

Regardless of the resources available to the campus under its control, a good campus plan requires working with the surrounding jurisdiction(s) to ensure that off-campus coordination is in place when a disaster strikes. FEMA has been providing funding for Disaster Resistant University program to increase resilience at America's campuses. (FEMA, 2003). During the 13th Annual FEMA Higher Education Conference a session was held to review the critical factors in campus emergency planning among a group of educators and campus leaders from around the nation. Their insights and suggestions, documented in this report, provide a rich resource for campus emergency planners.

Stakeholders:

As has been noted, campuses do not exist in a vacuum. They are within the boundaries of municipalities, counties or parishes, and states. Each of these entities has emergency plans that include the territory in which the college is located. Therefore, when campuses undertake to write an emergency plan the leaders need to be aware of all the stakeholders who are involved in emergency management for that area. There are immediate stakeholders on the campus. Faculty, staff and students are most directly involved. Among off-campus constituencies, parents are the biggest stakeholders in campus safety. All campuses have alumni and community supporters who can be part of the emergency response and recovery plans. Some colleges have investors whose interests have to be factored into emergency plans.

Campus groups that need to be included in emergency planning are campus law enforcement/security, campus administration, faculty senate, staff labor unions and student association. Plans must be made for closing the campus for a time, and consideration should be given to establishment of MOUs with other colleges for students to continue their education if damage will cause a closure exceeding a few weeks. This can be important in the case of a fire or hazardous materials accident which might deny certain courses access to laboratory facilities, as well as during campus-wide disasters. The role of faculty and staff in the response and recovery period should be included in labor agreements. In some state higher education systems the employees are all disaster service workers, as in California, requiring everyone drawing a pay check from a public source to continue to report to work during the disaster response and recovery phase and be assigned to do work under the Incident Command System.

Utilities, plant operations, buildings and grounds staffs all have important disaster planning roles. A decision must be made regarding who will be responsible for authorizing re-entry into campus buildings that have been affected by a disaster or emergency. Campuses with engineering programs might consider making arrangements with graduate students and faculty to assist with damage assessment, especially for buildings needed for shelter and medical care. Some pre-disaster training and certification might be needed to enable them to perform this function without taking on significant personal liability. The shipping and receiving/warehouse group and the mail room need to be included in an emergency plan. Emergency response and recovery may depend on the ability to receive goods that have been ordered. They are a special focus of concern during periods of campus unrest and bomb threats, as they may be the first to recognize a suspicious package and prevent its delivery.

Student health and medical services must be included in the emergency plan. A focus on health developed during the concern for an outbreak of an infectious disease such as swine flue and avian flu. Epidemic precautions were widely practiced on college campuses. However, rapid onset disasters like earthquakes and tornadoes require a level of preparedness for dealing with traumatic injuries and orthopedic injuries in the immediate aftermath of a natural hazard event when access to normal community emergency room services may be delayed. Campuses with nursing schools, pre-med programs and public health programs should consider post-disaster roles for students and faculty in these programs as a means of augmenting community resources for the campus. Schools interested in developing an in-house disaster medical response capability should consult their legal advisor, insurance carrier and state health department for guidance on training, qualifications and certifications needed for these personnel. While first aid trained non-medical personnel may have liability protection when rendering first aid under the Good Samaritan law, licensed medical personnel can only act within their scope of practice, which needs to be documented in advance.

Other means of augmenting campus medical capability might include MOUs with neighboring private practice physician groups and clinics. Recognize that in most states all licensed medical personnel with hospital privileges are required to report to the hospital to join the response, but if access to the hospital is not available due to infrastructure damage, a campus health center might be able to benefit from neighboring physicians' skills. Students who have registered with the Disability Resource Center may have additional rights to campus-based medical assistance. For example, students who require regular prescription medications may store their supplies at the student health service, and plans must be made to ensure that they will have access to their medications after a disaster. Students with dialysis needs should arrange with their medical provider for back-up equipment to be kept in student health for disaster use. Students with other disabilities should be encouraged to work with their medical care providers to create disability-specific disaster kits to augment the standard water/medications/food kit.

Feeding services and housing services are critical not only for resident students, but in a disaster also for commuter students, faculty and staff. Campuses with a full meal program may already have some stockpiles of food, but plans should be made for acquiring more food during a disaster to support the expected size of the disaster-based campus community. Housing staff, especially resident assistants, should be represented on the emergency planning committee. They can advise on dorm capacity and alternative locations for disaster housing for commuters, or alternative housing for residents of damaged dormitories.

Consideration should also be given to how housing for students with mobility issues is managed. If the elevator is not working (power outages, post-earthquake, for example) or is unsafe to use (in a fire, for example) how will students with mobility impairment evacuate the building? Have paid staff been assigned to assist these individuals to safety using special chairs, or using the individual's own equipment? For liability reasons the campus cannot count on student volunteers to assist people with disabilities unless they have been trained and have signed an agreement to provide the service. Students with their own live-in personal assistants need to ensure that the assistant has been trained in evacuation techniques and has signed an agreement to stay with the student until he or she is safely evacuated. These agreements should be on file with the university's Disability Resource Center. The best solution may be to house students with mobility limitations on the first floor of residential buildings to ensure ease of evacuation.

Community emergency services need to be part of the campus emergency planning process. Local police probably already have a relationship with campus police for mutual aid and campus/town border issues. Fire departments and emergency medical services may have a contractual relationship with the campus, or may see the campus as just another large business. In a large community a college campus is unlikely to get special treatment, but in small community where the campus is a major employer the campus may have a specific response plan within the community's overall emergency plan. Planning with the municipality's emergency management office will ensure that the community plan and the campus plan are coordinated, especially where the campus is relying on community emergency response agencies. Community political leaders will also be concerned about the effectiveness of the campus emergency plan, as success or failure of campus emergency response can reflect on the adjacent community and damage or enhance its reputation. For example, the high profile crimes on the University of Florida campus at Gainesville had a negative effect on the reputation for safety of the community of Gainesville.

In addition to fire department-based emergency medical responders, the campus may rely on private ambulance service providers, off-campus clinics and the Public Health Department, especially for contagious disease testing and management, notably tuberculosis and sexually transmitted diseases. In a disaster, management of safe drinking water and food preparation sanitation will also require the assistance of public health department sanitarians.

A college is also a business with vendors and contractors who need to be considered, including many community businesses that provide goods and services to the campus. A business representative can provide some insight into business continuity planning and reasonable expectations for service after a disaster. Pre-disaster MOUs may be prudent to ensure the best available response from local contractors and vendors. Campuses served by mass transit should include a representative in their planning, as they have resources to assist with evacuation, sheltering, movement of emergency responders and transportation of people with disabilities and mobility impairment.

Campus neighbors are also an important part of emergency planning. These may include neighborhood associations, chambers of commerce, and adjacent businesses. Campus neighbors may include K through 12 schools, nursing homes, hospitals, senior citizen housing developments and even shopping malls, as at Stanford University. Each of these entities will have an emergency plan for its residents, tenants or customers. Coordinating plans is important, so that one entity is not expecting to evacuate into another entity that plans to empty its facility. For example, neighborhood associations adjacent to campus may view the campus open spaces as post-earthquake shelter areas, while the campus envisions evacuating its constituents off campus to get away from potential building collapses or hazardous materials releases. Conversely, campuses may be counting on those open spaces for housing for their own commuter faculty, staff and students who are trapped at school by the failure of local roads. For example, in the San Francisco Bay Area the Association of Bay Area Governments (ABAG) estimates that 1700 road segments will be damaged, making driving anywhere difficult for hours to days (ABAG, 2003), so San Jose State University expects to use all open space on campus to house its own populations of 29,000 non-resident students and almost 2,000 faculty and staff who live elsewhere.

Most campuses create an emergency plan within the public safety services, often without much interaction with other elements of the campus community. Local municipal leaders from emergency management and emergency responders are seldom included in the planning, but the information from an exchange of views can be most enlightening. For example, San Jose State University reviewed its emergency plan in preparation for the state-wide Golden Guardian exercise in 2007, expecting the City of San Jose Fire Department to send rescue and hazardous materials teams to the campus as a top priority after an earthquake. It should be noted that the campus is a six square block state enclave within the city's downtown. While it generates a certain amount of business and sales tax through student, faculty and staff activities, the university pays no real estate taxes to the city. An exchange of e-mails with Fire Chief Darryl Von Raesfeld revealed that the city's emergency response plan called for the campus to be treated as just another downtown business, with response priorities set after the event based on triage of the events occurring simultaneously. This resulted in a realization of the importance of enhancing campus CERT training and creating additional on-campus response capability. Such an exchange of information regarding response protocols and priorities in a disaster could help to better shape the campus emergency plan to take into account the actual expectations of the surrounding community.

Resources

Resources to respond to an event should be listed in advance, and each should be contacted to ensure that the campus is part of that entity's emergency response plan. Water is the most important post-disaster resource. Emergency planners need to determine what bottled water is stored on campus for normal sales to students, staff and faculty, how much bulk bottled water is stored by departments for their internal break room supply, and how much clean potable water is in the boilers in the utility and heating plant.

Swimming pool water should never be used for drinking, although it may be used for sanitation, bathing and laundry if necessary. Other sources of bulk and bottled drinking water need to be planned for in advance to ensure that every entity in town is not counting on the same water bottling plant for supplies. Pre-disaster open purchase orders or other contracts should be made to ensure the campus' position on the supplier's priority list. Students, faculty and staff should be encouraged to make a car/desk/dorm kit with a minimum 3-day supply of drinking water at one gallon per person per day. Once purchased the stock should be rotated at least annually, or through regular consumption and replacement.

Medications are the next critical resource of emergency planning. Students taking medication regularly may have a supply in the room or stored at the campus health center. Will this supply be adequate for 5-24 hour days? What medications could health center staff prescribe and provide during an emergency for students with diabetes, seizure disorders, mental health needs or circulatory system problems? What medications could be prescribed and provided to staff and faculty during a disaster? Are staff and faculty told to carry a 5-24 hour day supply of their essential medications?Provide public education on personal support supplies to students, faculty and staff at least annually. Determine what supplies the university would routinely have, where they are stored, and whether it is likely that they would be accessible after a disaster. Determine what supplies could be obtained locally and make a formal arrangement for their delivery based on specific triggers in case communications systems are down. What supplies will the campus facilities and public safety staff need to carry on their duties during disasters: barricades, rope, tarps, generators and fuel for generators for food storage and lighting.

Facilities Development and Operations should ensure that all construction contracts include a clause requiring the contractor to relinquish control of the assigned personnel and equipment to the campus management section chief in a disaster.

The contractor will then receive a no penalty extension to the contract equal to the period of disaster work, and be paid for the use of personnel and equipment at the contracted price. Local police, fire, emergency medical services, emergency management staff and mass transit are important to campus response, but may be delayed by higher priority community events. Local utilities are also an essential resource for campus response and recovery efforts. These resources should not be written into the campus emergency plan until there has been a conversation with the public agency regarding the likelihood of their availability to the campus, and how long a delay should be expected before their arrival. Many campus personnel attending the session at the Higher Education Conference noted that tight budgets force them to rely heavily on community resources, which may not actually be available.

The campus public safety personnel then need to design a plan to replace or augment community resources in a disaster. The importance of CERT on campus has been discussed above. The campus has other human resources that may become emergency assets. There may be students on campus who are in the military, are veterans, or are police or fire personnel who might be able to assist if they were present when the disaster occurred, and were not legally required to report for duty elsewhere. Neighborhood associations might develop MOUs with the campus for mutual assistance during a disaster. Students with disabilities and health concerns may require special assistance during a disaster, some of which was discussed above. Community-based non-governmental organizations may be able to provide pre-event training to students, staff and resident assistants to help them help the special needs community. Sign language interpreters, hearing aid batteries, Braille and recorded materials may all be available through local NGOs in a disaster.

Campuses with a residential population will have different challenges than large commuter schools. Campuses in small towns will have different resources available than campuses in urban metropolises with staff and faculty commuting 50 miles and more to work. Each campus emergency plan resource list needs to consider the specific needs of the campus community it will serve: natural hazards, threat analysis for the campus, resident and non-residents students, faculty and staff.

Incident Command System:

Homeland Security Presidential Directive 5 (HSPD-5) mandated the use of the National Incident Management System (NIMS) for all agencies that receive federal preparedness funding from any federal agency. (Bush, 2003) NIMS includes the use of the Incident Command System (ICS) in the field, and a similar management system for the emergency operations center (EOC) that makes strategic decisions for the organization. California and Virginia have both adapted the field version of ICS to the strategic work of the EOC, although to date there is no specifically mandated system for managing EOCs under NIMS. In California it was developed in 1993 and is called the Standardized Emergency Management System. (California, 2007a).

Campus emergency response will have a field level tactical response, usually led by the security or law enforcement element of the campus. It will follow ICS. There will also be an EOC whose job is to management the event strategically, including managing resources, information, and coordination with other agencies and elements during the disaster. This is usually led by the senior administrative vice president, and supported by senior personnel from Facilities, Housing, Contracting and other business elements of the campus. A discussion of campus emergency management, including a complete set of EOC NIMS/SEMS position checklists for a campus EOC, are available in bioth.pdf and Word in The Role of Transportation in Campus

Emergency Planning (Edwards and Goodrich, 2009). At the FEMA Higher Education Conference the session that focused on campus emergency planning reviewed the roles of the campus emergency management team and suggested key elements that need to be included in campus planning considerations, along with the basic checklist elements. The session began with a review of the ICS/SEMS/NIMS system, and key elements in the formal system. Participant comments were intended to augment the material- including the specific checklists- available in the Campus Emergency Planning monograph cited above.

The Management Section is responsible for the overall management of the emergency. The Management Section Chief authorizes the opening of the EOC, and then meets with each Section Chief to create the Action Plan for the EOC, setting goals and objectives for each action period. This group coordinates with the university hierarchy (Policy Group usually consisting of the president and his cabinet), with campus elements through representation in the working sections of the EOC, and with student organizations and the local jurisdiction through the Liaison Officer. Session participants emphasized the importance of limiting EOC access to those assigned to EOC positions, noting the negative effect of "meddling" on getting decisions made and work done. One of the earliest requests for direction may be the relationship between the campus and its neighbors. Will the campus be closed for security reasons? Will campus facilities be open to neighbors needing shelter? Will feeding stations be for university ID card holders only or also for the community? If it is a public university, would this be considered a "gift of public funds?"

The Management Section has supporting Command Staff members who assist the decision maker. For example, the university's legal adviser should also be in the EOC to advise the Management Section Chief on liability and legal issues related to emergency management decisions. Questions like whether to shelter or feed people from the neighborhood have legal implications. Decisions on campus closures and tuition reimbursements also require legal advice. The emergency services coordinator, usually the day-to-day emergency plan manager, is responsible to ensure that the EOC is set up properly, and that all communications resources are working properly. This person must be knowledgeable about all elements of the EOC's functioning, and serve as a resource for all the other people working in the EOC.

The Liaison Officer needs to pay special attention to coordinating with both students and parents regarding the functionality and safety of the campus, and plans for maintaining classes and feeding and housing functions during the disaster. To that end he or she should maintain up-to-date contact information for students and parents in the EOC kit. This information must be updated at least once a semester. A Safety Officer is responsible for the overall emotional temperature within the EOC as people have to work together under stress. Participants noted the importance of having a strong person in this role, one who is willing to intervene when an EOC staff member needs to take a break or go home. Keeping everyone to the shift change schedule and ensuring that healthy food is available at meal times are important practical activities discussed in the monograph.

A Security Officer may be needed to ensure that only assigned EOC staff are permitted into the EOC. It is crucial to provide a secure location for the management of the disaster. A uniformed security guard or police officer is best for this role. The Public Information Officer (PIO) has one of the most important jobs in the EOC. Internally the PIO needs to ensure that the Policy Group gets copies of all media releases before they are sent to the media to ensure that the campus leadership knows what is happening in a timely fashion. (Note that they are briefed at the beginning of each new Action Period, but as events unfold media releases may update those briefings.)

Campuses may have students or staff with limited English capabilities, and who come from community sub-units that may have little or no English skills. Therefore it is important to determine ahead of time which languages are needed for press releases to keep the larger community informed, and for public education materials for on-campus students and staff. For example, at San Jose State University many of the janitorial staff members have Spanish as their first language, so radio announcements on campus status and personal preparedness material need to be available for them in Spanish. The PIO needs to have made arrangements with a native Spanish speaker to be interviewed by radio and television reporters, and with bi-lingual students or faculty to translate written materials into culturally and educationally appropriate Spanish for the campus and wider community. Before the disaster the PIO needs to take a census of language needs among students, staff and parents, and ensure that those needs are met during a disaster. At a minimum, pre-made written messages for the major threats should be available in the EOC PIO kit in all the appropriate languages, so that the PIO can just fill in appropriate information in a few blanks until linguistics support is available.

In a large disaster the community may establish a Joint Information Center (JIC). This will include PIOs from all the major agencies involved in the emergency response. The purpose is to ensure that timely messages are generated for the public, and that accurate and coordinated information is released to the media. (FEMA, n.d.b.) The PIO needs to ensure that a representative is trained to go to the JIC to represent the campus, and that that person has appropriate, working communications capability, whether cell phone, satellite phone or amateur radio, to ensure coordination with the EOC as events unfold. Students are used to social networking through communications. The PIO should ensure that someone on the PIO team is capable of disseminating information by way of text messages, the university's website, and even Twitter if campus policy permits. This means that at least one staff member per shift must be proficient in each of these technologies. Arrangements for battery operated portable electronic devices to support social networking and chargers for them should be part of the EOC plan. Remember that every car in the campus parking lot is a potential charging station for electronic devices, hand held radios and laptop computers.

Tactical management and decisions belong to the Operations Section. Typically the head of security or the campus Chief of Police would be the Operations Section Chief. This group supports the Incident Commander in the field and communicates directly with him or her to determine what support or resources are needed. The section also communicates with the community emergency response organization to ensure a coordinated response. Requests for police and fire mutual aid, emergency medical services support and environmental health support would come from this section.

Because the campus may not have a large responder group, appropriate faculty members might become EOC staff. Faculty teaching public policy, criminal justice, emergency management and homeland security would be good selections to augment university staff. (Economics and finance professors, planning professors. journalism professors and others might be useful for staffing other sections of the EOC.) Branches within the Operations Section could include Care and Shelter staffed by Campus Housing, Construction and Engineering staffed by the physical plant leadership, and medical and mental health staffed by the Student Health Service, who would coordinate campus services.

The Housing staff should have a list of housing assignments to facilitate search and rescue and space reassignments when facilities are damaged. Housing may be required to develop shelter for commuter students, staff and faculty who are on campus when the disaster occurs. Sheltering may also be needed for EOC staff members so they can stay on campus between shifts. They should coordinate with Logistics to acquire needed tents, trailers or motel accommodations.

The Planning/Intelligence Section Chief is responsible for information collection and dissemination, as well as for beginning the recovery process. Damage assessment is the first concern of the P/I Section, using campus resources when permitted, as described earlier. The Damage Assessment Branch collects safety information for all buildings that may be re-occupied after the disaster, within legal limitations. The building surveys must be documented, preferably with the ATC-20 or state OES-provided forms. At a minimum a campus map should be used with a code to indicate building conditions, photos should be taken of all buildings as they are surveyed, both outside and inside if entry is safe. This information is collected and shared with the community to become part of its official disaster report to the state government. The Situation Status Branch maintains a log of all EOC activities, and records significant changes to the campus condition. These documents are essential for reimbursement from insurance companies and possible FEMA funding. The P/I Section Documentation Branch makes maps, collects information on weather conditions from NOAA, sunrise and sunset times, and other factors that may affect the way the disaster is managed in the field. They also fill out all forms required by the state and federal government, and collect information on all aspects of the campus status. They make the organization chart for each Action Period in the EOC showing who is filling each EOC position. They use whatever system is in place to communicate this information to the higher levels of government. In California there is a web-based system called Response Information Management System (RIMS) that is used to communicate required information using forced-entry forms. (California, 2007b) Paper forms can also be faxed or scanned and sent via ham radio packet systems if the internet is not functioning.

The P/I Section Chief is also responsible to organize the Action Planning Briefing at the end of each Action Period where all section chiefs meet with the Management Section Chief. The P/I chief documents the decisions that the Management Section Chief makes and creates the written Action Plan for the next operational period. The plan states the strategic level goals and objectives, the length of the next Action Period, and includes the updated organization chart, and any supporting plans and materials. The plan becomes part of the EOC log, and is given to each section chief, and the Operations Chief verbally communicates it to the Incident Commander in the field.

Recovery begins while the disaster is still unfolding. The Recovery Unit of the P/I Section is responsible to ensure that plans are made for business recovery and restoration of services. They coordinate with the Logistics Contracting Unit to get debris removal contractors, roll out boxes for disposal of damaged personal goods, rental items like sanitation facilities and tents or trailers for residential use.

The Logistics Section includes the Information Technology Unit that manages computers and all forms of communications hardware. They are responsible for surveying the availability of communications on and off campus, and to using substitutes when traditional methods are unavailable. They issue satellite phones and cell phones from the EOC cache, and coordinate RACES operators for positions without working phones.

The Contracting/Procurement Unit acquired goods and services needed for response and recovery. This unit must maintain a list of sources of goods and services, standing purchase orders, and 24-hour contact information for all vendors that are likely to be needed for emergency response and immediate recovery. Camping goods stores, motels and hotels, restaurants with 24-hour service and delivery service, generator rental companies, information technology sales, fuel delivery companies and food and water vendors would be the top items on the list. While ideally the EOC is in a building with a generator installed, powering the EOC for the duration of the disaster response is one of the first items that must be arranged for. They need to know who takes purchase orders, who will deliver the needed goods, and which organizations have 24-hour delivery. They need contact information for Facilities maintenance personnel in case the EOC needs repairs. All the vendors on the list should be contacted in advance, and the 24-hour contact information must be updated every month. A staging area, staffed by Contracting Unit staff, has to be set up to receive and document all the goods as they arrive

One of the first actions of the Contracting/Procurement unit is to arrange for food and drinks for EOC workers, and to support the field response as needed. Water and non-caffeine drinks should be stored in advance to make EOC opening easier. Some simple snacks should also be stored in the EOC, such as granola bars, crackers and canned juice packed fruit. After the EOC is opened a regular schedule of meals should be established when possible. Alternatively, employees should be trained to bring their own emergency kit items with them to the EOC.

The Personnel Branch is responsible for staffing the EOC, and for assisting the field Incident Commander with personnel call backs and mutual aid requests to fill field roles, unless there is a law enforcement unit in the Operations Section that is handling law enforcement mutual aid. This branch is usually staffed by campus Human Resources staff members who bring their employee rosters and contact information with them to the EOC. These people not only tell people when to come to work to help with the emergency, but they also call people to tell them to stay home if they are not part of the emergency response. They are responsible to notify all paid faculty and staff if the campus is closed, for example.

The Finance/Administration Section is focused on tracking expenses, paying bills, and collecting data needed for reimbursement. Personnel from Accounting and Payroll usually staff the F/A Section. They will coordinate with Logistics staff regarding allowable expenses. At the end of the event they will coordinate the collection of photos, logs and maps from other EOC staff to facilitate reimbursement for losses and expenses from insurance companies, and from the state and FEMA as the law allows. In advance of the event the F/A leadership should contact the state emergency management agency to determine eligibility for reimbursement for emergency response costs and emergency repairs, and receive training on documentation requirements for the reimbursement programs.

The F/A staff will track costs, overtime and over budget expenses. They are responsible to keep the Management Section Chief informed of any over budget costs for emergency work, and to recommend ways to reallocate budgeted funds to pay for the emergency.

The F/A staff has to maintain the financial systems of the campus. Payroll must be issued and benefits must be paid for. In a disaster the employees may need to use sick leave or vacation to manage family disaster issues, and medical care for disaster-related injuries, so they are counting on the campus to keep its bills paid. Many staff members may receive paychecks rather than having direct deposit. Without power it may be difficult to get the paychecks printed and delivered to employees. One alternative is to have an outside source print and mail the paychecks to employees. This could be done through a commercial contract or through a mutual aid agreement with another university. Either system requires updating every payday. After Katrina, Northrup Grumman paid their employees through Western Union. Employees could go to any Western Union office with a driver's license and an employee ID card from Northrup and get cash. (Edwards, 2006)

F/A should maintain a cache of paper forms and checks to allow them to process all requests for payment independent of the computer-based system, in case power is out or the computer equipment on campus is damaged. Petty cash in small bills and change may be needed to pay for food from small vendors, as ATMs and check cashing equipment are likely to be damaged and out of service for the first several days.

4

Transport Infrastructure

INTRODUCTION

Transport infrastructure policies create impacts not only to the transportation system, but also to social, economic and environmental systems. As individuals and as part of society, we perceive immediate and/or gradual changes such as travel times and costs. Obviously, other non-immediately perceived changes occur and affect planning objectives and goals. This may affect the analysis of impacts and how effectively the objectives ofthe Land Transport New Zealand Act and New Zealand's Land Transport Strategy (an integrated, safe, responsive and sustainable land transport system) are met throughout planning actions.

A review of the scientific and technical literature on impacts of transport infrastructure polices reveals that most methods and techniques rely on the static and unidirectional representation of transportation-activity systems interactions (NCHRP, 2002; Hartgen, 2003, Wray et al, 2000; Weisbrod and Beckwith, 1992; Boarnet and Haughwout, 2000; Sanchez, 2004). Most studies perform a posteriori analysis of impacts, *i.e.* after the implementation of a policy, observed and expected changes are compared in order to evaluate the policy's efficiency and effectiveness. This approach is based on the assumption that the impacts of transport infrastructure policies are solely the result of immediate changes in the transport and activity systems.

Despite the widespread adoption of this approach, there has been growing criticism regarding the lack of comprehensive databases and evaluation methods that help planners in identifying the true dimension of the impacts (Greiving and Wegener, 2001; Badoe and Miller, 2000; Hirschman and Henderson, 1990; Prastacos, 1986; Shaw and Xin, 2003)In contrast to previous research efforts, this paper explores the dynamic linkages between transport and land use as well as the implications to socio-economic and environmental changes due to transport infrastructure policies. This approach follows recent research initiatives (Rodrigue, 1997; Miller and Demestky, 2000) that have incorporated the time dimension in the analysis and modeling of dynamic interactions between transportation and activity systems. A framework for analyzing impacts of transport infrastructure policies under a holistic approach is presented. The paper is divided into five sections. After this introduction section, a brief review is presented on theoretical concepts of impact analysis. The third section introduces the proposed impact assessment framework. The fourth summarizes a case study of the impacts in a highway project. Finally, the main findings and potential lessons learnt that could be assimilated in New Zealand are presented in the fifth section.

CURRENT PRACTICE IN IMPACT ANALYSIS

Activity-transportation systems interactions are critical in assessing impacts of transport infrastructure policies. Activity System is defined as the totality of social, economic, political and other transactions taking place over space and time in a particular region.

On the other hand, the transportation system consists of the facilities and services that allow flows throughout a specific area, region, country, etc. Aiming to provide conditions to perform movement of people and goods, transportation is a factor of activity location, which in turn is dependant on with specific land uses. Manheim (1980) represents these interactions as shown in Figure in which both systems are directly related and consequently any modification in one may affect the other.

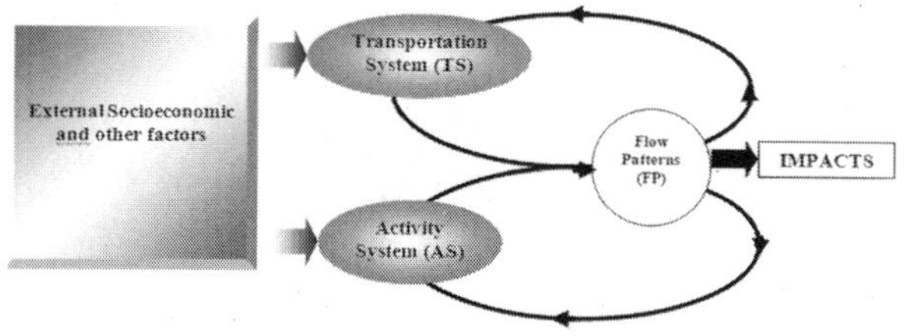

Fig. Interaction of Transportation and Activity Systems
Adapted from Manheim (1980)

As an example of these interactions, consider a highway segment that is upgraded in its service level at given year 0. Over time, the improvements to the highway segment create changes to surrounding environment. As shown in Figure, land use (*e.g.* more densification and diversification due to increase in traffic), environment (*e.g.* more noise and air pollution as traffic density increases), demographics (*e.g.* migration of population due to new activities) and economy (*e.g.* more economic activities due to decrease in travel costs) gradually change throughout the region. Subsequently, in year 25, further interactions and changes are observed.

Impact analysis focuses in specifying the effects of any changes on the activity and transportation systems (Meyer and Miller, 1984; Manheim, 1980). Based upon the current transportation system characteristics and future predictions of activities, many methods and techniques to analyse and evaluate impacts are found in the scientific and technical literature. They concentrate in estimating monetary benefits and costs in order to assess relative merits of planning actions (Dickey, 1983).

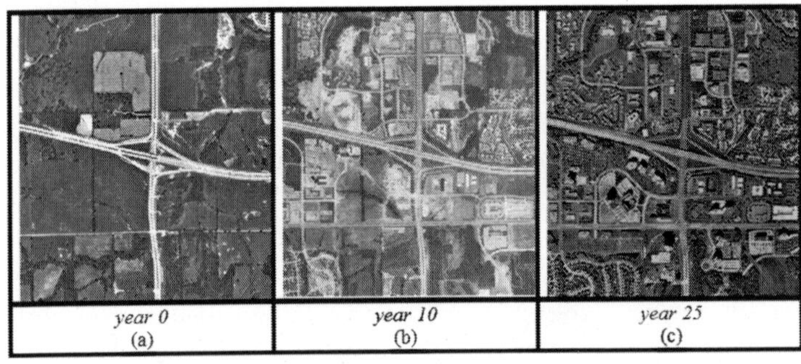

Fig. Spatial and Temporal Interactions between Transportation and Activity Systems

However, despite the diversity of methods and techniques applied to analyse and evaluate impacts, static analysis and evaluation of impacts is mostly observed (Shaw and Xin, 2003). The analysis and evaluation methods and techniques concentrate on just one side of the whole planning process, which is either the transportation or the activity system as observed in researches. Furthermore, recent research (Badoe and Miller, 2000; Ryan, 1999; Hirschaman and Henderson, 1990; Prastacos, 1986; Shaw and Xin, 2003) indicates that adopting current state-of-art analysis and evaluation methods and techniques may lead to inaccurate and/or incomplete assessment of the impacts of transport policies. Consequently, the analysis and evaluation may ignore dynamic effects and changes in the surrounding environment.

IMPACT ASSESSMENT FRAMEWORK OF TRANSPORT INFRASTRUCTURE POLICIES

This assessment framework comprises a set of steps in order to incorporate and integrate dynamically a wide range of transportation and activity systems affecting the analysis of infrastructure policies. The main theoretical assumption is:

For example, be a transport infrastructure asset, such as highway segment, which is under analysis as part of a planning process. One on hand, Infrastructure Policies (IP) may alter the physical and operational characteristics of the existing assets and/or recommend construction of new infrastructure. On the other hand, Socioeconomic Profile (SP) may also contribute to affect the activity system. The combined effect of IP and SP changes result in Impacts (IM) such as variations in travel patterns, traffic composition and volume, travel time and speed, accidents, air quality and noise, population, land use, etc. In a temporal perspective schematically, future impacts (IMt+n) would be the outcome of:

- All previous impacts (IM^{t}, IM^{t+1}; .... IM^{t+n-1}); and

- All previous infrastructure policies (IP^t, IP^{t+1}; .... IP^{t+n-1}); and
- The over time socioeconomic profile (SC^t, SC^{t+1}; .... SC^{t+n-1}); and
- The temporal characteristics of the transportation system ((TS^t, TS^{t+1}; .... TS^{t+n-1}) and the activity system (AS^t, AS^{t+1}; ....; AS^{t+n-1}).

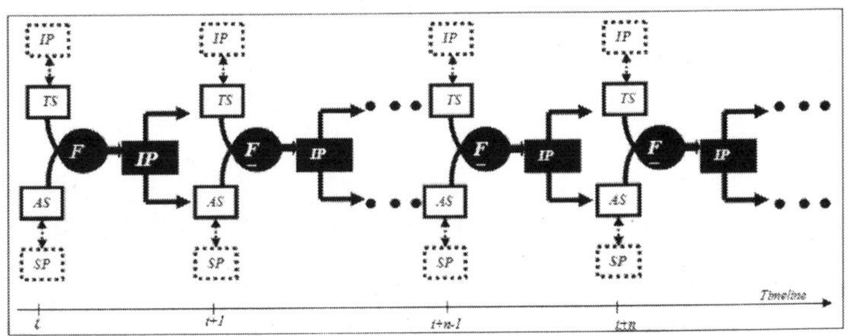

Based upon this definition, the impact analysis is conducted as per the following steps:

- Define the scope of the analysis (asset type and characteristics) and the intervening elements (Transportation and Activity Systems);
- Build a spatial-temporal database of study area, including the transport infrastructure and the activity system;
- Compute spatial-temporal changes for each intervening element and respective attributes (characteristics);
- Perform cross-reference analysis overlaying observed changes (step 3) from multiple layers of spatial-temporal data; and
- Identify short, medium and long term impacts from the cross reference analysis, based on cause and consequence tendencies.

Case Study

Similarly to other countries, New Zealand transport policies and projects are analysed in order to assess their effectiveness in terms of meeting the community needs and governmental goals. More specifically, the Land Transport New Zealand Act and the New Zealand's Land Transport Strategy require information about the transportation and activity systems interactions over time and space. Currently, this information is obtained from the examination of the transportation system's direct impacts (*e.g.* travel time, accidents, vehicle operational cost, etc). However, this is not exactly in accordance to the complex and dynamic nature of the interactions occurring in the surrounding regions of transport infrastructure and services. Therefore, state-of-the-art planning tools are often considered as inaccurate to predict the impacts of transportation policies and projects.

It is envisaged that the framework presented in section three can play a significant role in addressing the observed lack of planning tools. Above all, the framework could contribute in creating information about the land-use and transport system interactions occurring over time Therefore, the framework is the first initiative to create better information than currently available. This is achieved through the consideration of a wide context in which transport infrastructure and service affect the community and surrounding land use. Consequently, the framework would contribute to expand project evaluation practices, which are mostly focused on direct impacts of the transportation system.

Aiming to understand the effects of an infrastructure policy, the proposed framework is applied in a case study. This attempts to show the potential of the proposed framework in order to analyse the impacts of various policy interventions in the main highway corridor named Presidente Dutra Highway (PDH) in Brazil. In the last 50 years, PDH has contributed to the development of the Southeast region of Brazil. Constructed in 1951, PDH provides the means for goods and people movement trade between Rio de Janeiro and Sao Paulo and the rest of the country. At time of construction, PDH was a high quality two-lane highway, with 405 kilometers of extension and 115 special works of art (turnpikes, viaducts, bridges and lower crossings).

Due to transportation demand increase, as a consequence of economy growth in the 1960's, the government upgraded PDH to a four-lane highway in 1969. The investment contributed to the reduction of 40 per cent of the transport cost between Sao Paulo and Rio de Janeiro and induced more traffic flow between both cities (Silva, 2000). However, in the 1980's, a period of recession and shrinkage in infrastructure investments contributed to the deterioration of PDH's service levels due to reductions in maintenance and construction funding. In 1996, PDH was transferred to the private sector through a concession contract in order to provide a high quality highway with modern services and support the economic development. Thus, 5 toll plazas were installed and tolls fees are charged from users.

Considering this background, the results of each step of the assessment framework are presented below:

- Step 1: Taking for reference the boundaries of the Rio de Janeiro and Sao Paulo states, the study region was delimited. This region comprises 167 municipalities as shown in Figure.
- Step 2: For each municipality census and administrative data sets and physical and operational characteristics of the PDH segments were obtained. Main data items were: Population, Gross Domestic Product (GDP), Urban, Industrial and Rural land uses of the municipalities in the study region; Traffic Volume and Number of accidents of each segment of the PDH. Using this data set, a georeferenced database was created in a GIS software (GeoMedia Professional), as shown in Figure.
- Step 3: The main spatial-temporal changes are summarized in Table.
- Step 4: A major development tendency and impact of the infrastructure policy interventions is the formation of Development Centers (DC) and Development Gaps (DG) in the PDH region. DC

and DG are the result of completely different growth patterns. Significant changes in population, socioeconomic characteristics, transportation systems and land use patterns occurred in coincident areas clustering blocks of cities. In between the DC, DG (cities with low levels of changes) are also observed. Figures 5, 6 and 7 show the spatial distribution of the DC and DG.

All DC have the same growth tendency for all factors in the 1970-2000 period. The only exception is the number of accidents, which experienced a considerable reduction in the last 20 years. Another interesting aspect is the modification of the land use patterns. In almost all cities of the PDH region, there was a sharp tendency of industrialization and urbanization along the highway. On the other hand, three main gaps that have been affected with low change levels are observed. Despite the proximity to PDH, population, socioeconomic and the transportation characteristics have changed less than other cities in the DG. The interesting and common aspect of these cities is that the land use (especially rural) has changed considerably, which can indicate modification of the development pattern in the near future.

The process of DC and DG formation is the outcome of the combined effects of each intervention. The initial intervention (Capacity upgrade with federal government funding) produced general changes in socioeconomic, population, land use and transportation features modified almost at the same level. Originally, the PDH region had very limited number of activities, but this changed in the 1970's due to a period of industrial activity booming, which is reflected in generalized increase in traffic flows, accidents and economic development. The adopted policy of limited maintenance (1980-1990 period) associated with the economic and political crisis produced considerable deceleration of growth changes. Although DC 1 and 2 still have achieved high values in terms of population, socioeconomic and land use characteristics changes, they were less intense than before. On the other hand, transportation characteristics experienced a huge decrease in growth changes, in which DC 1 and 3 dropped considerably.

The upgrade and maintenance through PPP and toll charging scheme (1990-2000) produced a slight increase in terms of population, socioeconomic and land use characteristics in all DC. As for the transportation changes, DC 1, 2 and 4 decreased, while DC 3 suffered a huge increase in all respects. In the 1990-2000 period, the introduction of the toll system (concession of the PDH) altered the traffic behavior, which may have resulted in the modification of development patterns. However, after the implementation of this policy, this development pattern was altered. In areas nearby the toll plazas, drastic changes in terms of growth were observed. This has generated the transference of development previously observed in DC 2 and 3 to the DC 1 and 4. Moreover, the travel demand in between the toll gate plazas increased considerably, which indicates that short-distance trips became more cost effective because there was no additional charging for those trips. Simultaneously, socioeconomic indicators demonstrated a sharp decrease in activities especially in DC 2 and 3, which is mainly due to the increase in the perceived additional costs (toll fees).

There is a strong relationship amongst travel pattern and the changes in population, socioeconomic, land use. It is clear that before the infrastructure policy (concession of the PDH) there was a growth tendency without limits for all activity-highway system factors. Nearby the metropolitan areas, the low level of transportation costs (provided a high standard of accessibility) and land use (or property) values are possibly the main conditions encouraging this development pattern and creating conditions for increasing in goods and people movement as well as a large variety of activities.

- Step 5: Developments tendencies and impacts are expressed in terms of the dynamic interactions between population, socioeconomic characteristics, transportation system and land use changes over time and space. The policy interventions are grouped into three main categories, namely: capacity upgrade with federal government funding; minimum maintenance with federal government funding; and recuperation, upgrade and maintenance through PPP and toll charging scheme. Trade-offs between policy interventions and tendencies/impacts are examined in order to identify cause-consequence relationships that are used to assess the implementation of the tolling system.

Long term prospects considering the current location of toll plazas indicate that future developments will reach saturation. Probably, there will be the relocation of new activities just beyond the boundaries of the toll gate areas in terms of activities and transportation capacity in the areas of DC 2 and 3, because transportation costs and land use values will increase considerably in the areas of the DC 1 and 4.

Conclusions

A framework for analyzing impacts of transport infrastructure policies under a holistic approach was presented. This framework attempted to incorporate the time dimension of dynamic interactions between transportation and activity systems and linkages and implications to socio-economic and environmental changes due to transport infrastructure policies.

A major outcome of the case study is that a static analysis of impacts generated different results to those obtained by applying a temporal assessment framework. Hence, it was observed that infrastructure policies have created short, medium and long-term impacts of extreme complexity. These impacts are the result of a continuous development cycle in which transport infrastructure policies trigger changes in land use patterns (activities, location and density), socioeconomic and demographic indicators as well as operational conditions of the highway. Based on the understanding of the development cycle in the PDH region, the recent implementation of the tolling system through a PPP scheme created impacts in terms of initial limitations to development in all respects due to the increase in direct (user perceived) transport costs. The charging system caused saturation (land use and transportation) of those areas in between toll plazas.

The proposed framework and the case study findings may contribute to the evaluation of highway projects in New Zealand. The spatial-temporal analysis of impacts may be used to expand the range of factors that are currently considered in the Land Transport NZ's Project Evaluation Manual. Even though the proposed framework does not include all possible intervening factors, the consideration of a broad range of variables may help in understanding the development cycles and the long-term impacts of infrastructure projects. This understanding will be essential to assess how infrastructure policies will meet the objective/goals of the Land Transport Strategy. Moreover, the proposed framework would help decision makers to conduct short, medium and long term monitoring of impact of their decisions. Further studies should be conducted in order to enrich the knowledge about the impacts of infrastructure polices. A suggestion list for future research is presented below:

- Any change in the infrastructure policy should be made considering all the aspects related to economic, political and social issues and not only the highway operational reality. Also it is advisable that changes should be conducted according to a national strategic plan, which defines objectives, steps and efficiency measures that reflect on welfare of the nation;
- During the execution of the case study, several difficulties were faced in organizing the database for the study region. It is imperative that planning agencies at all levels dedicate more effort to creating a unified data standard and collection process that will eventually contribute to form a data warehouse. This will be crucial to developing further studies and reaching new findings; and
- Creation of instruments to predict development scenarios based on initial estimates of population and economic growth in the region of the highway project. Such an instrument has to consider the spatial and temporal dynamic of all the involved variables. Currently, such an approach is not observed in the literature of transport-land use modeling.

NETWORK PLANNING AND INFRASTRUCTURE DESIGN

Cycling is wonderfully liberating and enervating in the right environment: it is difficult and unpleasant if the environment is not suitable. Programmes of activity to promote the values and benefits of cycling encompass elements of what are sometimes called the four 'Es': engineering; education, enforcement and encouragement. While such a range of interventions, perhaps particularly with a strong emphasis on cycling skills training, may assist in promoting cycling, it is not easy to achieve significant uptake in cycle use without the development of a physical environment right and proper for the particular characteristics of cycle users.

The appropriate balance between promotional activities and infrastructure provision to maximise levels of cycling, however, remains elusive. For example, in a review of evidence from the UK Cycling Demonstration Towns project, it has only been possible to conclude that the sustained investment had an effect, and it has not been possible to determine more closely defined effects from specific interventions (Sloman et al., 2009). A longer historical view suggests an interesting interweaving of infrastructure provision and cultural development of cycling. What remains clear, however, is that the planning, design and construction of any route that cycle traffic is expected to use must be directed towards making the route fit for its purpose. Strategies for promoting infrastructure for cycling have sometimes only considered the specification of engineering upgrades to particular localised areas or supposed problematic features of the existing road network. However, a strategy that is more likely to succeed will consider a whole settlement or a whole network, and such a strategy, as with the construction of motorway networks in many western countries, for example, may take decades to complete successfully.

Although not the focus of this chapter, it is worth discussing the nature of cycle users and the range of non-infrastructure interventions that are available. Road users in charge of motorised vehicles are adult or near adult, have usually had to pass a test after having been trained, and may be subject to health checks to avoid them driving after the onset of dotage. By contrast, cycle users exist from an age after which they can balance on two wheels, which typically may be four years old.

The health benefits of cycling allow for a user to continue cycling, if he or she so wishes, easily into a tenth decade. Training of cycle users, where it exists and where it is taken up, is typically less rigorous than that for motorised vehicle users, and in any event not compulsory. A cycle user provides the motive power and the distribution of speed of cycle users ranges from the lowest speed possible to retain balance (around 12 km/h to avoid significant lateral movement, CROW, 2006) to speeds exceeding 40 km/h for fast commuters (Parkin and Rotheram, 2010). The kinematic envelope of a moving bicycle and rider combination is typically taken as 1.0 metre (DfT, 2008) and the rider's body is exposed to the environment through which he or she is passing. Overall, cycle users are heterogeneous in terms of their physical characteristics, have a wide range of cognitive abilities and are, owing to their relative exposure, vulnerable in hostile environments.

Non-infrastructure interventions used by those promoting cycling in countries with low cycle use relate to the enhancement of an individual's knowledge and skill (for example cycle training, provision of information through maps and signs), and behaviour (for example relating to safety, road sharing and road traffic violations including enforcement), and also extend to celebration of cycling (for example through rides and festivals, marketing and promotion including tourism promotion). In addition, other actions relating to good planning and governance have also been used including the monitoring of use and opinion, the establishment of reference groups of cycle users, and professional training in provision for cycling. Necessary infrastructure developments that are not intrinsically part of the transport network include the provision of high quality cycle parking and the development of appropriate carriage and integration arrangements with public transport.

There has been an increase in the amount of guidance published by and for road authorities relating to infrastructure design for cycle traffic and Parkin (2010) provides an evaluative review on the most prominent of these. The review discusses levels of service for cycle networks and particularly considers the issue of the effort of the rider, and includes a discussion on design speed. We take the opportunity here to extend the discussion to consider issues surrounding the contention that exists between protagonists of on-road provision and off-road provision (the integrationists and segregationists introduced by the authors of Chapters 2, 4 and 5). To do this we use a framework based on an analysis of risk. Finally, and importantly, we critically assess the currently promulgated so called 'hierarchy of provision', which purports to guide planners and designers to specific types of action in an ordered way. We propose a distinctive new approach for planning the creation of any physical infrastructure needed to promote cycling based on an appropriate analysis of the needs of cycle traffic. We conclude the chapter with some rumination on the path to sustainability for transport through infrastructure provision for cycling.

PRINCIPLES FOR DESIGN

On the basis that the approach for designing routes for moving vehicles is well established in design manuals for highways, and is based around a design speed, it should be clear that the guiding principle for design for cycle traffic ought also to be that the bicycle is a vehicle capable of speed. For routes designed as transport corridors only for bicycle traffic, designers need to ensure that they apply appropriate principles of highway and traffic engineering in a similar way as they would apply those principles for motor traffic. Geometric features of the route alignment need to be designed according to the selected design speed, and these include curve radii in the vertical and horizontal planes, sight distances required for stopping and overtaking, and lengths of tapers for lateral movements within the route. In addition to the engineering essentials of highway and traffic engineering, however, the designer needs to be aware of the special characteristics of the bicycle and rider, which in turn influence the design characteristics appropriate for cycling infrastructure.

These are suitably defined by the Dutch (CROW, 2006) as being coherence, directness, attractiveness, safety and comfort. These guiding characteristics have been adopted by authors of guidance across the globe in varying forms. While also being important for all types of transport network, there is significant added weight placed on many of these attributes for the reason that the rider is very much part of the embodiment of the bicycle as a vehicle. Directness is more important than for other types of transport because of the added personal cost of effort, and perceived safety is paramount because of potential feelings of vulnerability when passing through certain types of environment. Relative to more sedentary forms of travel, comfort will already be compromised because of the nature of the vehicle and the balance required riding it, and so issues such as surface roughness become even more important.

There has been room for a significant amount of opinion amongst traffic engineering professionals and advocates around the specific 'working out' of traffic engineering relating to the bicycle. Some contrasting approaches include: provision of on-road cycle lanes; provision of segregated cycle paths in the road corridor, usually behind the kerb line; provision of cycle paths completely away from road corridors, *e.g.* through parks or utility corridors. 'Bicycle boulevards' and 'neighbourhood greenways' have been popular in North America, and the Dutch pioneered the 'woonerf' shared space concept in residential areas. More globally there are contentions about whether 'providing for cycling' is the same as 'providing cycle facilities', with lower speed limits and/or lower traffic volumes often being sufficient to create a cycle-friendly environment. Forester (2001) provides an exposition of the argument that there is little by way of provision that ought to be made for cycle traffic as distinct from other types of vehicle, although this relies on riders undertaking fast and sometimes necessarily assertive 'vehicular cycling' in relatively hostile environments in order then safely to co-exist with other road users.

These different viewpoints often arise because of the different perspective of the protagonist, who may either assume that a person cycling always wants to achieve the highest speeds possible and is comfortable mixing with motor traffic, or that cycle users are 'afraid' of traffic and wish to avoid motor traffic at all costs (for a fuller discussion of the role of the informed citizen as a protagonist, see Chapter 4). For authorities trying to provide for the widest cross-section of existing and potential cycle users, it can seem difficult to reconcile these conflicting viewpoints. In particular, much of the concern can often centre on the perceived quality and level of service that would be afforded by the alternative options of 'on-road' cycling and 'off-road' cycling. We argue that this dichotomy is false so long as designers adhere to the engineering principles of route design and the five Dutch characteristics for cycle networks. Clearly, a lot of the contention has been linked with the extremely poor consideration given to cycle traffic in some developed countries to date, and this can be associated with poor knowledge amongst traffic engineers of the characteristics of cycle traffic, comprising as it does of vehicles travelling at speed. Critical issues for design are: a smooth bound surface; suitable geometry for the design speed; suitable speed management of all traffic; suitably constrained gradients; adequate width; properly designed crossings and transitions from on-road to off-road; full and proper route maintenance.

For reasons connected with a well developed sense of urban design, and for reasons connected more with a desire to manage traffic appropriately, many urban areas have central zones that have had vehicle traffic excluded, *i.e.* they have been 'pedestrianised'. Allowing cycle use to penetrate into areas designed more specifically with pedestrians in mind (street malls, squares and so on) has been a contentious issue in some countries, but is relatively common in many parts of Northern Europe.

Frequently, the effect of prohibiting cycling in pedestrian areas is to 'force' cycle users onto longer, busier traffic routes, designed specifically for motor traffic management purposes. Davies et al. (1998) confirmed this by finding that many alternative routes involved high capacity roads, additional hazardous junctions, additional distance and the majority required cyclists to dismount at some stage. In many cases, there is adequate capacity in motor-free areas to safely cater for all non-motorised users. Trevelyan and Morgan (1993) analysed video and questionnaire responses from sites in England and Wales and examined conditions in other countries and found that people cycling respond to pedestrian density and modify their speed, dismount, and take other avoiding actions where necessary. Collisions between pedestrians and cyclists were very rarely generated in the areas studied; in fact only one pedestrian/cyclist collision was noted in fifteen site-years. This is supported by similar findings from German surveys with initial public reservations being significantly reduced after a year's experience, and evidence of the adaptation of cycling behaviour, including dismounting, when high densities of pedestrians were present (CROW 1993). There was no evidence that cyclists rode more quickly once legally allowed in pedestrian areas, and pedestrian-cyclist collisions were small in number and not too serious.

Overall, routes for cycle traffic without the presence of motor traffic can be advantageous where such routes allow for access much closer to the final destination, where they allow for a more direct route to be taken, and where they are designed to be attractive and comfortable for cycling use. However, the network offered for use by cycle traffic should be fit for purpose based on guiding characteristics of coherence, directness, attractiveness, safety and comfort. Before outlining an approach for developing networks suitable for cycle traffic, we review an area of discussion that arises frequently when considering cycle use, the question of risk to users.

The Question of Risk

Movement and risk are inextricably linked, and this linkage results from the high probability of collisions between moving objects in close proximity and the physics of momentum transfer and force created in collisions. Clearly, where human and animal life is involved in such collisions, the issue of risk becomes an ethical one, and of great significance to society as a whole. Road safety engineering has approached the subject by analysing data to assess and develop engineering interventions to assist in reducing the incidence and severity of collisions. Enforcement of appropriate behaviour plays a role (for example, the enforcement of speed limits), but more recently added to the armoury of road safety practitioners are encouragement (of a safety culture whereby individuals accept responsibility for their own and others safety) and education (through training and publicity) (ROSPA, 2003).

A network of roads and road users forms a system for movement, and the premise underlying interactions between users is that they 'drive on sight'. There is a distinction between the objective 'visibility' of a vehicle (for example car, or bicycle and rider) and the ability, for whatever reason, of road users to appropriately perceive other moving objects or people in the road. In countries with relatively low cycle use, like the United Kingdom and New Zealand, cultural emphasis has been placed on the responsibility of the person cycling to be 'visible', which simultaneously and implicitly reduces the perceived need of other road users to properly use their senses to 'perceive' the other vehicle or person. To some extent, a rebalancing has been recently attempted by Transport for London in a series of advertisements that suggest to the general public and road users in particular that 'the more you look for something, the more obvious it becomes'. Cycle users generally take up less road space, although they are frequently taller than most private motor vehicles. On the basis of their relatively smaller size, it could be argued that they are relatively less visible. However, a difference in relative visibility is not an argument that a 'lazy' road user should be able to make for not having perceived a person cycling: it comes back to the principle that the road network is offered on a 'drive on sight' basis.

The risk of a collision is related to absolute speed and the relative direction and relative speed between moving objects. An absolute increase in speed does not necessarily increase the risk of collisions if the road system is designed to minimise differences in the direction and relative speed of vehicles. For example, a system of motorways segregates directions of travel, provides appropriately angled merges and diverges at junctions and bans slower moving vehicles. However, in mixed traffic conditions often found in urban areas, an increase in speed results in a higher risk of collision and higher consequential damage because of the presence of many slower moving road users. Risks of collisions are enhanced by the relative differences in speed between road users, and failures to yield priority appropriately in accordance with the usual rules of the road. Behaviour and interactions between road users will vary and depend on the attitudes of the individuals involved, the road space available, the form of infrastructure and traffic management and environmental conditions, such as the degree of light and the weather.

Existing modern transport networks and traffic management regimes were often built with limited or no thought being given to cycle users. The provision of retro-fitted 'facilities' for cycle traffic has often been undertaken to attempt to address this issue. While these facilities may often ostensibly have the aim of reducing one or more perceived risks of cycling, they may also have other undesirable effects such as reducing priority for cycle users, or the speeds that they would otherwise wish to travel at. Adams (1995) differentiates between risks perceptible and controllable by the individual as a result of their actions (in his example, climbing a tree) and risks that are perceptible only with the help of science through an analysis of the evidence (for example, the prediction and control of diseases). He suggests that an individual will make decisions based on an inherent propensity to take risk, but influenced by rewards and experiences of losses due to accidents. On the other hand, governments take action based on an attempted analysis of the objective probabilities of 'risk and reward', and it is eminently debatable the extent to which reasonable decisions are made based on such 'corporate' analysis of risk on behalf of the individual (this is why debate continues about nuclear power and military action, for example).

Adams also points out that the management of risk will modify the risk and hence change behaviour. An important distinction is behaviour that changes a risk to oneself and behaviour that changes risk for another, for example driving with inappropriate speed for the condition may be relatively safe for oneself, but not for a third party. On this basis, it could be argued that a constraint placed on a driver as a result of a 'facility' for cycling (*e.g.* a narrower traffic lane) may in fact result in the driver seeking to compensate in some way and taking more risky action in order to produce the same reward (at the most banal, being 'home in time for tea'). Such compensation may have more significant detrimental consequences than if no perceived 'constraint' had been placed in his or her way.

Perceptions about risk are frequently quite different than an analysis of the accident record might suggest. For example, pedestrians may avoid crossing a road at a certain place, even though it may lie on a desire line, because of their perception of the hazard of crossing at that place. This reduction in pedestrian use will be reflected in fewer collisions involving pedestrians. Such feelings of insecurity are just as 'real' as the forces experienced in a collision. So far as cycling is concerned, a major issue is concerned with the conditions in which it is appropriate for cycle traffic to mix with motor traffic. A common misconception about off-carriageway routes is that they are inherently safer than their on-carriageway counterparts simply because of the absence of motor traffic. Other risk factors, such as other cycle users and pedestrians, and objects at the side of the route, still require cycle traffic to behave on motor traffic free routes in a disciplined way and to 'ride on sight'. People new to cycling in particular have a strong fear of collisions with motor traffic, so routes for cycle traffic off the carriageway are, seemingly, attractive to them. As evidence to support this, Kingham et al. (2011) found in New Zealand a preference for separated facilities (separate from both motor vehicles and pedestrians) across a sample of people interested in cycling but currently not regular utility cyclists.

The majority of cycle collisions do not involve motor vehicles: people fall off or hit objects for various reasons, and they also have many collisions on routes shared with pedestrians, dogs, and other people cycling. Munster et al. (2001) estimated from New Zealand hospital data that four times as many cyclists are injured from 'cycle-only' crashes on the carriageway or on footways and other routes than those involved in motor vehicle collisions (note that these data do not include off-road mountain-biking track accidents). When considering children specifically, Safekids (2007) concurs, and suggests that 90 per cent of New Zealand hospitalisations for bicycle-related injuries to children during 1999-2003 did not involve a motor vehicle. Similar findings have been found elsewhere (Moritz, 1998; Carlin et al., 1995; and also Franklin, 1999).

It is also worth observing that conflicts with motor vehicles may not be reduced by off-carriageway riding; cyclists will typically still have to cross side roads and driveways, where most conflicts occur. Forester (2001) points out that a key assumption for advocating off-road paths is that same-direction motor traffic is the greatest danger to cycling (*e.g.* being hit from behind). With American data, he showed that these types of collision made up only 1 per cent of all cycle collisions both on-road and off-road.

Road injury collision data for 2006-2010 in New Zealand shows that 58 per cent of reported urban cycle collisions are at intersections, with a further 19 per cent occurring at driveways; it is difficult to avoid those, even on a footway. Considering specifically collisions that could probably be avoided by cycling adjacent to, rather than in, the carriageway (*e.g.* overtaking, hit car door, rear-ended), fewer than 30 per cent of all on-road cycle collisions appear to be likely candidates.

In moving cyclists to a separate route, however, additional collision opportunities may be introduced, especially if the route is within the road corridor. Nearly 10 per cent of all reported cycle collisions in New Zealand note that the cyclist was (illegally) riding on the footway (sidewalk); with more than half occurring at driveways. Depending on the location, frequent conflicts with pedestrians are possible and there may be less perception and reaction time for conflicts that occur at driveways or side-roads. While the geometry and physics point towards such additional collision opportunities, the extent to which they may in fact manifest themselves in reality will be a function of the way that riders and drivers interact, and this is an artefact of behaviour.

A number of studies in North America have found that collision rate involvement when cycling on footways is higher than on the carriageway or on off-road cycle routes (Aultman-Hall and Hall, 1998; Moritz, 1997). An interesting finding by Aultman-Hall and Adams (1998) was that regular footway cyclists also had higher on-road crash rates than non-footway users. This raises the possibility that footway riders are either less confident or lack the skills and training of on-carriageway riders (although they did find that regular commuters had similar collision patterns on footways). The different problems faced at intersections compared with mid-block locations are highlighted by Danish research (Jensen 2008) that found that, while off-road cycle tracks were safer in general than their on-carriageway counterparts, they were less safe at intersections. SWOV (2010) therefore recommended that cycle tracks parallel to roads should either rejoin roads ahead of intersections or be taken further away to cross the side roads.

A concern remains amongst cycle users that a collision with a motor vehicle is more likely to lead to serious injuries, hence perhaps the preference to use the footway. Certainly most bicycle-related deaths involve a collision with a motor vehicle. Over the ten-year period 1998-2007, typically six out of every seven children killed in bicycle-related incidents in New Zealand resulted from a collision between the child and a motor vehicle. But, while moving the cyclist off the carriageway may result in a lower incidence of severe injuries, it may result in transferring injuries to pedestrians that are hit by cyclists (albeit rarely fatally). Aultman-Hall and Hall (1998) found that the likelihood of 'major' cyclist injury remained about 1.7 times greater on footways than carriageways.

By contrast, neighbours to off-road cycle routes sometimes express concern about crime, vandalism and litter, and these risks of public disorder may affect their property's value. A review of 300 off-road routes in North America by The Rails-to-Trails Conservancy (1998) suggested these misgivings were unjustified and the Trails and Greenways Clearinghouse (2003) identifies potentially significant economic benefit that can arise from popular new trails to small communities. A range of surveys suggest the majority of local residents consider off-road routes to be a useful additional local amenity and an advantage for, or at least a neutral effect on, property values and public safety. Sustrans (1999) note that visual appearance of the finished product will have an effect on the perceived and actual benefits and this fits in with the core Dutch characteristic of attractiveness. One of the problems with many existing off-carriageway facilities, at least in New Zealand and the United Kingdom, is that they have been poorly designed and maintained (if at all) for cycling. Indeed, a lot of the above research focused on cycling on existing roadside footways designed for pedestrians. The central point is that when routes for cycle traffic have been created off-carriageway, cycle traffic has been expected inappropriately to fit into an unsuitable environment for such traffic, with little or no design consideration having been given to the bicycle as a vehicle capable of speed, and this has contributed to the collision problem.

An important way of managing risk on the road network has been to ensure that every aspect of geometric design is related to the design speed. In order to manage the risk on any route that is deemed appropriate for cycle traffic, an appropriate design speed must be considered by the designers. Such a design speed is relevant whether the path of the cycle traffic is within a carriageway used by other motor vehicles, whether it is along a route dedicated to cycle traffic but intersecting with roads and pedestrian routes, or whether it is along a route that also serves pedestrians. In this latter case there are two possibilities: routes are divided into a carriageway for cycle traffic and a footway for pedestrians, or, in a manner akin to country lanes that have no footway, there may be a single shared surface. The design speed dictates forward visibility envelopes for stopping and overtaking, curves in the horizontal and vertical planes, and the tapers over which vehicles are expected to make lateral movements within the space available to them. The use of public roads is well codified in law and in user guidance, such as, in the United Kingdom for example, the Highway Code (DfT, 2007). In some limited instances it may be appropriate for routes to be shared by pedestrians and cycle traffic without the pedestrians being offered a footway adjacent to the carriageway dedicated to cycle traffic use. However, when remote from motor traffic, it would appear that the good sense promulgated in the guidance is not adhered to. In particular, the UK Highway Code suggests to pedestrians that they walk towards oncoming traffic on the right hand side of the road (the UK has a left hand rule of the road for driving) in order to face oncoming traffic. This rule is not widely promoted as a 'rule of the route' for routes that cater for both pedestrian and cycle traffic, but would, if implemented, provide a suitable codification for behaviour on such routes. Alternatively, many North American and Australasian shared pathways mark and sign the path like a road, with all path users in the same direction keeping to one side unless overtaking.

Milton Keynes is one of the so-called New Towns in the United Kingdom that was developed through the 1960s to the 1980s based on a concept of easy motor vehicle access and relatively low development densities. In the case of Milton Keynes, the transport network comprises a grid pattern of roads connecting at large roundabouts with cycle and pedestrian traffic banished to separate 'redways' that dip beneath the road network through underpasses. Franklin (1999) identifies reasons for a high accident record on the redways as being poor design and very poor user discipline, for example the flouting of basic practices such as cycling on the left (mirroring the left hand driving rule of the road) and not using lights at night. He also notes increased conflict between pedestrians and cycle users as a cause for concern, and this suggests that where routes are designed for speedy cycling progress they should, in a similar manner as for 'normal' roads, be kerb separated from adjacent footways. Despite the network of cycle routes, cycling has never achieved a significant proportion of use and this is likely to be as a result of the town having been designed around and for the car in particular (rather than the bicycle), a lack of priority for cycle traffic and a general lack of cultural support for cycling.

A range of studies demonstrate reduced risk associated with greater numbers of bicycles within the traffic mix which have spawned the idea of 'safety in numbers'. In an excellent commentary, Bhatia and Wier (2011) point out that the literature consistently supports non-linear models in which increases in injury collision frequency are proportionally less than the increase in the volume of the road user being considered. However, they suggest that greater clarity of argument is required, particularly where public policy intervention is being considered, in order properly to recognise the difference between reduced individual risk and overall total number of injury accidents, which, for changes in mode share to more benign modes of a low magnitude, will still result in an overall increase in injury accidents. They also point out that confounding may occur (for example safer environments will increase both numbers and safety) and that adequate evidence of the causal direction is lacking (in other words increased safety could be producing increased numbers) in current 'safety in numbers' studies. They do suggest, however, that plausible causal mechanisms could exist, such as more users being more visible, more users forming larger groups, which in turn may have some sort of collective vigilance. Overall they ask why streets should not be designed for adequate safety regardless of the number of users. Now having gained an appreciation of the issue of risk and cycling, we turn our attention to the sort of networks that would support cycle use.

NETWORKS FOR CYCLE TRAFFIC

This section provides guidance for developing comprehensive transport networks suitable for cycle traffic. Individual links and junctions in the network may or may not happen to be coincident with routes over which motor traffic also has a right to pass, but it is likely that they are coincident in the majority of cases. The guidance offers direction to those promoting cycling, providing transport, and planning and undertaking spatial planning and land development. Taken together they may be regarded as a set of values to help position cycling appropriately within the whole transport system for a twenty first century sustainable society.

Despite the ubiquitously quoted engineering requirements and characteristics, their operationalisation has frequently been left wanting. A favourite approach has been to create a hierarchy of provision of treatments to existing routes (for example IHT, 1996; DfT, 2004), which may first attempt to reduce motor traffic volumes, followed by reducing speeds, followed by intersection treatment and traffic management, followed by reallocation of carriageway space, followed by the introduction of specific facilities such as additional bridges. This was introduced in Chapter 4.

Parkin (2010) has argued that such a hierarchical approach to implementation leads the designer down the path of making adjustments to the existing network at the individual route and corridor level without initially considering properly the higher-level introductory processes of understanding demand, and then providing for that demand with a suitably constructed network defined by engineering parameters and the Dutch network characteristics. The hierarchy of provision is not helpful in outlining the processes of route and network planning that must precede scheme implementation.

Virtually all cycle trip ends will be on the road network, that is to say most trip ends will also be accessible by motor traffic. This may be quite acceptable if one end is a residential road, but, based on the planning and engineering approaches of the latter part of the twentieth century, the shopping or business end of the journey may be surrounded by high volume, high speed roads which may not be conducive to an attractive and comfortable environment in which to cycle. This leads to the realisation that, unless constructing a completely new town or suburb, the introduction of suitable networks for cycling needs to be retrofitted into an existing well established and frequently heavily used road network. Retrofitting may require no action at all: the existing road routes may be fully suitable and attractive for cycling. However, they may on the other hand benefit from specific forms of traffic management that may provide more appropriate movement space for cycle traffic as a distinct type of vehicle within the traffic mix. We suggest below actions in a procedure for the implementation of infrastructure to assist in promoting cycling.

- Spatial planning
- Demand modelling
- Traffic and speed management and the creation of enhanced permeability
- Construction of links to complete a permeable network
- Integration with public transport

Action 1 would normally precede or be undertaken in association with Action 2 and these are the usual steps in a local authority's approach to managing the physical development of its jurisdiction. Actions 3 to 5 are interconnected and complementary and may be undertaken in parallel with each other. Not all actions may be as desirable and necessary as other actions, and this will depend on the nature of the areas being considered.

Spatial Planning

The rate of change of land use varies between cities and countries across the world, but is generally slower paced in older established communities and nearer the centre of larger communities. However, spatial planning determines the geographical end points of journeys and this geographical relation, and the intervening transport network, help determine the demand for transport and the modes and routes used. The nature of the relationship between land use and transport demand has been discussed in Chapter 5. Cervero et al. (2009) summarise the five relevant spatial planning attributes in an interestingly alliterative style as being:

- Density;
- Destination accessibility;
- Design;
- Distance to public transport; and
- Diversity.

In a consideration of integrated transport, Hickman et al. (2010) expand these five attributes to eleven themes including:

- Settlement size (with larger settlements offering more mixed land uses and less need to travel);
- Location of major growth areas;
- The nature of the strategic transport network for medium and long distance travel;
- Density of land use;
- The balance between employment opportunities and housing;
- The accessibility to key facilities;
- Development site location;
- The extent of a mixture of land uses;
- Neighbourhood design and street layout;
- Approaches to travel demand management; and
- Policy on car parking.

Owing to the effort required, cycling is better suited to short journeys. Evidence from the UK National Travel Survey (DfT, 2011, Table NTS0308), for example, shows that 85 per cent of bicycle trips are for distances up to 5 miles, and this contrasts with 56 per cent of trips by car as a driver being up to 5 miles in length. The themes identified by Hickman et al. that act as particular reinforcement for cycling include higher land use densities, mixtures of uses, and good juxtaposition of housing with jobs and services. In addition, cycling could be further supported through appropriate management of travel demand for other modes and suitable policies to constrain car parking (and at the same time provide sufficient bicycle parking). In much larger settlements where distances are greater, and for journeys between settlements, planning can be used to encourage cycling as an access and egress mode from public transport.

At a more strategic level, the planning framework should be used to provide and protect corridors for cycling related infrastructure. Frequently, routes for cycle traffic have been created opportunistically based on the availability of, for example, disused railway corridors. While these may be appropriate for leisure purposes, they are of limited value for transport purposes unless they happen to lie on desire lines that satisfy a demand for movement. New Zealand guidance (LTSA, 2004) points out that a cycle route separate from a motor traffic route may only be practical when planning new suburbs and townships. Such practicability is, though, in essence a political question. Historically, it has been deemed appropriate to use planning and highway laws for the compulsory purchase and construction of highways for motor traffic. In more recent times, there is evidence that these powers have been used for marginal road widening for junction improvements linked with bus priority. Ultimately, a decision about the nature and layout of a transport network in an urban area is related to public policy. It is implicit that adequate and timely transport planning needs to be undertaken to ensure that routes with a suitable level of accessibility and which may be constructed to suitable standards for cycle traffic are identified and protected through the appropriate legal processes.

Strategies to promote cycling do not exist in isolation from other policies and actions. The success of cycling strategies is greatly influenced by what is done elsewhere; other policies and programmes must be consistent with the promotion of cycling. This integrated approach also helps to emphasise the fact that the cycling strategy is not an 'add-on', but an integral part of the activities of road authorities and other agencies. Therefore, agencies need to review and implement other pro-cycling policies. These include road projects that should take full account of cycling needs and the use of other routes and corridors, which may or may not be in existing public open space but may have value as transport corridors for cycle traffic. An example of developing integrated thinking has emerged in London and is evident through the requirement that those engaged in designing or maintaining the street network are all well versed in the London Cycle Design Standards (TfL, 2005).

The ability to develop appropriate approaches for provision for cycle traffic will be influenced by variations in local conditions, for example in terms of topography, climate, compactness of land use and historic settlement structure, and social and cultural attitudes to cycling. The extent and quality of infrastructure may also be constrained by local economic and political priorities in the allocation of often scarce urban realm and road space and other transport policies and investments such as public transport investments. To ensure continued progress, spatial and transport planning requires long term commitment to a policy direction and sustained commitment from a range of public authorities, private organisations and non-governmental organisations at different levels (local, regional and national). These organisations may include local employers and health, education and leisure agencies. A further constraint on the ability to achieve success may be the availability of appropriately qualified and experienced professional staff and this deficiency may to some extent be overcome by appropriate training and professional development.

Demand Modelling

While spatial planning is an on-going process punctuated by a particular timescale (usually dictated by a legislative need for renewed land use planning policies on perhaps a quinquennial basis), the planning of specific infrastructure such as a transport improvement, will happen at a particular point in time and be related to a particular project as part of a scheme of investment. Demand modelling is therefore an activity that would be undertaken at a specific point in time relative to a planned programme of investment as a result of a policy initiative.

Erstwhile provision of infrastructure for transport has frequently been caricatured as being a process of 'predict and provide' with roads 'filling up' with motor traffic because of rising car ownership and use virtually as soon as they have been constructed. Based on the experiences in the late 1980s and early 1990s in the UK, a detailed analysis of this phenomenon of seemingly relentless growth was investigated (DoT, 1994) and evidence of traffic being 'induced' as a consequence of the provision of infrastructure was revealed. This resulted in significant changes to the way the forecasting demand was undertaken. So far as cycling is concerned, the approach presently adopted may perhaps be described as 'provide and promote', and this makes the bold assumption that whatever is being provided is fit for purpose. Infrastructure interventions to promote cycling have not always resulted in a hoped for increase in use of cycling. This is linked with the extent and quality of provision, but also with the reluctance of a population to switch from other modes while the provision for cycle traffic remains, in their eyes, incomplete.

Cycle demand modelling is still in relative infancy and Bamberg in Chapter provides an overview of a series of emerging methods for modelling choice for cycling. Methods that have been used for investigating the level of use of cycling have otherwise been related very closely to a study of its relationship with the extent of provision of infrastructure. For example, Nelson and Allen (1997) examined cycle pathway length and its relation with cycle commuting in eighteen cities in the USA. Each additional mile of cycle pathway per 100,000 population was associated with a 0.07 per cent increase in cycle commuting. However, based on the data being cross-sectional in nature, it was not possible to infer a cause-and-effect relationship.

Extending the work to thirty five large cities, Dill and Carr (2003) found that the percentage of people commuting to work by bicycle correlated significantly with various cycle infrastructure variables, but not with any other transportation, environmental or demographic variables. The strongest correlation was with the number of on-road bicycle lanes per square mile. Each additional mile of bicycle lane per square mile was associated with a 1.0 per cent increase in the proportion of workers commuting by bicycle. (It should be noted that such an increase in bicycle lanes is significant as the cities in the study averaged only one third of a mile of bicycle lane per square mile). Again, no cause-and-effect relationship may be claimed, but it does imply that some commuters will adopt cycling if the infrastructure is appropriate and this has been discussed more fully in Chapter 5.

Katz (1996) used a stated preference approach to estimate an elasticity of demand of 0.6 for cycling based on changes in the proportion of a journey served by some sort of bicycle 'facility'. This finding suggests that a 25 per cent increase in facilities would increase the numbers of cycle commuters by approximately 15 per cent (25 per cent × 0.6). Parkin et al. (2008) used UK census data and revealed that the proportion of people who cycle to work was most affected by the geographical factors of hilliness (elasticity -0.893), rainfall (-0.665) and temperature (+0.703) with the proportion of bicycle route that is off-road displaying a low elasticity of +0.049.

The point is made that revealed data contrasts with larger elasticities usually found in models constructed from stated preference data. McDonald et al. (2007) examined ten case studies in New Zealand and derived predictive equations for cycle numbers on on-road and off-road facilities. The equations were based on existing cycle mode share and growth in the local district and, in the case of off-road facilities, the adjacent motor vehicle volumes. While an estimation method was developed, the researchers acknowledged that further case studies would help improve the model. The field of estimating cycle use is still evolving, and there are now some useful guides available to help predict likely future cycle numbers using a variety of techniques (including FHWA, 1999; Katz, 2001 and DfT, 2010). The methods range from simple comparisons of similar facilities previously built, through the use of elasticities, to full scale trip demand models. In addition, as already mentioned, Bamberg in Chapter 9 introduces a wider range of socio-psychological models and Zuidgeest and Brussel in Chapter 8 describe geographically based techniques for defining and overcoming barriers through cycle network construction.

Traffic and Speed Management and the Creation of Enhanced Network Permeability

The geography of the transport an urban area is defined by not only the layout of the road network, but also by the way traffic speeds and volumes on the road network are managed through area wide strategic traffic management. Historically traffic management has been implemented usually only with a view to 'smoothing the flow' of motor traffic and what is now required is the creation of systems for easy movement for all classes of traffic which recognises the very different nature of different types of vehicle, in particular the bicycle.

Future area wide and strategic schemes will need to be thoroughly operationally appraised using fully specified transport models and junction analyses such that different scenarios of layout and priority may then be compared. Additional regulation by the creation of, for example, one way streets and banned turns may be one outcome, but also it may be appropriate to construct new links within the network specifically for cycle traffic, for example, in the same way that hitherto perhaps 'ring roads' or 'relief roads' have been built for motor traffic in the past. New links are discussed further in the section below.

While specific traffic management measures may be necessary for the size and volume of motor traffic, these measures may not be necessary for cycle traffic, which is much more agile in nature and has much less impact on the urban environment than motor traffic. An alternative network, more advantageous to cycle traffic should always be considered, and may be regarded as being 'interleaved with' the network as defined for motor traffic. This will create enhanced permeability (sometimes referred to as 'filtered permeability', see Melia, 2007) for cycle traffic. The creation of such preferential networks would rely on traffic regulation and also, again, in some instances, on the creation of specific new links.

It is technically quite feasible to manage the physical infrastructure of the highway in a variety of ways which provide priority for different classifications of vehicle. The greater difficulty is in ensuring that the priority is being appropriately considered for cycle traffic from a political point of view: funding only follows for such traffic management schemes with such support.

As discussed in Chapter 4, advocacy for cycling has played an important role in shaping thinking, even if sometimes it has established antagonisms between 'users' and 'providers'. The consciousness of the importance of creating civilised urban environments through speed management has been significantly raised in recent times by international campaigning around the theme. 20s Plenty For Us suggests that 'total', that is town-wide or settlement-wide, zones with speeds of 20mph (30km/h) would create a new 'societal norm' for speeds in urban areas and reduce the need for expensive physical measures to 'calm' traffic to appropriate speeds. Ownership of such an approach, it is suggested, would be enhanced because most drivers would benefit in their home street, a consistent message that 20 mph is 'plenty' where people live can be easily promulgated and accepted, and that, being based on a democratic process, an additional collective community commitment to road danger reduction can be created. Evidence, for example, from the early adopting City of Portsmouth in the UK suggests a 22 per cent reduction in road collision casualty numbers (20splentyforus, 2011) and such town specific evidence is supported by a recent review of the literature (Reid and Adams, 2011), which points to speed limits as being an important risk factor in multi-vehicle collisions involving cycle traffic.

Interestingly, based on a conception that the European Commission's 2010 road safety strategy paper was too vague, timid and inadequate, Members of the European Parliament Transport and Tourism Committee provided democratic support for reduced speed limits in urban areas by endorsing a report that calls for a 30km/h speed limit in all residential areas and on single-carriageway roads without cycle tracks. Specifically they see this as a road safety measure to help cut the number of children aged less than 14 years old killed on roads by 60 per cent and those seriously injured by 40 per cent (European Parliament, 2011). There is hence, therefore, also evidence of political will in support of technical strategies to create more hospitable transport environments.

Even without a corresponding speed limit reduction, traffic calming on its own can provide useful safety benefits. Mao & Koorey (2010) investigated nineteen streets in Christchurch, New Zealand, that had been reconstructed with various vertical and horizontal traffic calming devices. They found that the average collision frequency decreased by 17 per cent, despite the general trend of increasing collision numbers on Christchurch's local roads.

The Dutch guiding characteristic of coherence addresses the point that everyone capable of cycling should be able to use the road network to access all of their desired destinations by bicycle. A significant challenge is the often difficult issue of maintaining attractive and comfortable sections of the network at locations where it intersects with or runs parallel with roads that carry significant volumes of motor traffic. Therefore, consideration needs to be given to treatments that ensure adequate provision for cycle traffic. Such provision may include a degree of lateral separation from motor traffic to reduce the impact on the cycle user of the negative impacts of moving traffic, with air pollution and noise perhaps being the most important amongst these. It may also include specific measures to manage conflict at junctions, such as marked space for cycle traffic through to the provision of distinct waiting areas (for example, dedicated approach lanes, or advanced stop lines to create a 'box' area) or times during a signal cycle only for cycle traffic.

Construction of Links to Complete a Network

It is not self-evident that the existing layout of a road network is ideal for cycle traffic. While some very old established routes may lie on obvious desire lines between specific settlements, there are many newer roads that have been constructed specifically with motor traffic management in mind, for example inner 'ring' roads. Bearing in mind the shorter average journey lengths and the need for directness for cycle journeys, these routes may or may not lie on useful desire lines for cycle traffic. As with management of the network for motor traffic where new build is deemed appropriate, there is the possibility that new build specifically to cater for cycle traffic could be required. This may connect, for example, housing estates via an area, such as a green swathe, which may easily be able to accommodate cycle traffic, but would be detrimentally affected by motor traffic. The construction of such additional links may require crossings of geographical barriers such as rivers and railways (where it may be too costly to provide a motor-vehicle carrying structure) and this affords civic society the opportunity of adding value to their neighbourhoods through the elegant design of walking and cycling bridges.

As mentioned in the section on spatial planning, it may be necessary to protect possible routes for cycle traffic with appropriate planning policies, and this needs political will and a desire to match the aspiration with funding. Conversely, such schemes may catch the imagination of local politicians and become 'flagship' schemes which may garner wide public support and the opportunity for kudos and the all important future vote of constituents.

Integration with Public Transport

Cycling coupled with public transport can provide a powerful alternative to the private motor vehicle. The door-to-door capability of cycling can be combined with the long-distance range and speed of public transport (especially where dedicated public transport corridors are available). Public transport can also provide a useful backup for cycling in bad weather. To achieve this, public transport systems need to allow for both good-quality cycle parking at stops and stations (DfT, 2008). Debrezion et al. (2008) found in the Dutch context that the availability of bicycle parking has a positive effect on the choice of railway station for the start of a journey. Ensor et al. (2010) determined that the provision of cycle and public transport integration in six major urban New Zealand centres would collectively generate more than 1.7 million trips per annum, with individual cities having benefit-cost ratios between 2 and 10.

Rietveld (2000a) found in The Netherlands a higher percentage use of the bicycle as an access mode to the railway than for the share of bicycle use more generally. However, a lower percentage uses the bicycle as an egress mode and he suggests policy to provide more secure parking provision at the destination end of a journey for a second bicycle, public bicycle schemes or better facilities for the carriage of bicycles on trains. The carriage of bicycles on public transport may be possible depending on the nature and capacity of public transport vehicles and the loading characteristics. It is usually not possible for bicycles to be carried on peak time trains on commuter routes, but limited carriage may be possible at off-peak times. Long-distance passenger services provide a desirable service for cycle tourists. Rietveld (2000b) further contends that overall journey speed, including the access and egress legs of the journey, is the main criterion that railway companies should consider when assessing rail journeys against alternatives.

THE PATH TO SUSTAINABILITY

The chapter has reviewed the principles for design of cycle networks and considered the question of risk, It has also argued for a more sophisticated and comprehensive procedure for planning and designing for cycle traffic than is offered by a hierarchy of provision, and which begins with spatial planning, adopts appropriate demand modelling, and then carries out schemes based on appropriate new build where required, and traffic management at a strategic and area wide level.

A common problem for politicians is the need for relatively quick results from investment. Any less-than-significant or visible change to the number of people cycling in an area may lead to calls for a moratorium on investment in cycling. Cycling England, a relatively short-lived quasi-autonomous non-governmental organisation established between 2005 and 2011, was aware of this issue and successfully concentrated its resources on the Cycling Demonstration Towns where concerted action did produce results and hence provided the evidence to seek additional funding for an expansion of the programme.

Koorey (2003) considered the state of the cycling network in Christchurch, New Zealand where considerable investment had occurred over the previous decade. While the rhetoric may sound impressive to some that over 60 km of streets had cycle lanes, with another 60 km of off-road cycle paths being available, it has to be remembered that there were approximately 1500 km of urban roads overall in Christchurch. There remained about 200 km of busy arterial routes that many potential cyclists would avoid and there were in many localities cases where there are limited opportunities to cross routes carrying significant volumes of motor traffic. The cycle 'network' was no more than a series of 'islands' isolated from each other. Clearly, if the general road network were comprised of such an isolated grouping of sub-networks, it would have a significant detrimental impact on motor vehicle trips. (One of the authors asks his students to consider how the volume of traffic on the M6 motorway may be affected if a link is missing say between junctions 16 and 17.)

Growth in cycling numbers will only become significant when networks suitable for a wide range of cycle users reach completeness and maturity. The fact that often the most difficult (and yet crucial) elements of cycle networks are left until last because of financial or political pressures only exacerbates this problem. The 'easy' and less necessary and influential components are often completed first, but it is generally not these aspects that are limiting the potential for growth in cycling numbers. More than minimal investment of time and resources is required to overcome these problems.

Many of the perceived problems with cycle facilities, both on-road and off-road, have been due to inadequate design or maintenance standards rather than the choice of facility. This of course can (and should) be resolved. Design that minimises risk is of concern for all transport networks and there are safety issues with both on-road and off-road cycle 'facilities'. Obviously it is desirable that all types of provision for all types of traffic are made as safe reasonably practical.

Even with appropriate design, the general public or local politicians may still develop incorrect perceptions about the relative merits of some 'treatments' designed to improve cycling. Education campaigns may be useful to inform these parties of the true value of such treatments in terms of safety and level of service. Some of the best solutions to promote cycling may not involve 'remedial' facilities, but involve the overall presentation of a network as suitable for use merely through good promotion.

5

Urban Transportation Strategy and Development

INTRODUCTION

Urban transportation policy in China is mostly driven by addressing traffic problems because traffic congestion has become one of the most serious urban problems in many large Chinese cities. For example, in central Beijing (within the third ring expressway), the average traffic speed was 45km/hour in 1994, 33km/hour in 1995, 20 km/hour in 1996, and 12 km/hour in 2003. On some arterial roads, the speed has dropped to 7 km/hour. During the rush hours, about 20 per cent of roads and intersections are in total gridlock and the traffic speed is less than 5 km/hour. The average speed of buses in major central cities has dropped from 30-35 km/hour in the 1950s, to 20-25km/hour in the 1970s, 14-20km/hour in the 1980s, to 10-15km/hour in the 1990s. In 2003, the average bus speed was about 9.2 km/hour in Beijing, and 10 km/hour in Shanghai in 2004.

These traffic problems, caused mainly by rising resident incomes, rapid urbanization, and the national policy to encourage automobile ownership in order to stimulate economic growth, have created other serious problems that affect urban growth, economic development, and the quality of life of residents. For example, in 2003, more than 40 per cent of commuters in Beijing spent more than an hour to commute to work. Only 5.5 per cent of workers took less than 20 minutes to commute. Traffic also causes grave problems for the urban poor as we will discuss later in the paper.

To address these increasing traffic problems, the Chinese government has focused mainly on increasing the supply of road infrastructure by expanding road systems and developing rapid transit systems like subway, light rail, rapid bus transit, and even the magnetic levitation (Maglev) system. For example, the City of Beijing drafted the "Beijing Transportation Development Framework" in 2004, and the City of Shanghai developed the "Shanghai Metropolitan Transport White Paper" in 2002. Both plans focus mainly on increasing road and rail capacity and improving transportation infrastructure.

By adopting this supply-based, capacity building transportation development strategy, the cities wish to alleviate traffic congestion and modernize the urban road infrastructure. It is this desire to show off the achievement of municipal government by modernizing urban infrastructure that motivates most city governments to build more roads and rapid transit systems. Therefore, modernizing urban infrastructure becomes the de facto guiding principle in most urban transportation plans and development strategies in China.

As a result, we see the development of expressways, subway, light rail, and rapid bus transit systems consistently at the top of transportation plans in most Chinese cities. At the same time, traditional non-motorized modes and bus systems are always at the bottom of the list in the transportation plans and development strategies. Even worse, walking and biking are discouraged in some cities not only because they are considered to impede automobile traffic flow, but also because they are considered a relic of the past, and derided as inferior transportation modes. The neglect of walking, biking, and busing in transportation planning policies will have severe negative impacts on the accessibility needs of the urban poor, since these non-motorized travel modes and buses are used by the majority of urban residents, particularly the low-income and the urban poor.

The urban poor are a special group of people with very low incomes and very limited mobility. Due to increased urban unemployment and growing numbers of poor farmers migrating to urban areas, the population of this group has risen over the years. These urban poor rely on non-motorized modes and buses to get around in their daily life, to access work places, medical, and other essential services.

Despite a large amount of investment in transportation systems, the accessibility of the urban poor in large Chinese cities has not improved. Walking and biking, the most common modes of transportation for the urban poor have been made even more difficult. In some cities, biking is even prohibited. Furthermore, redevelopment of older portions of the cities forces many residents, often low-income, to move to the suburbs. This movement makes it more difficult or even impossible for low-income residents to commute to work by walking or biking. Big investments in rapid transit systems like subway, light rail, or bus rapid transit can help. But due to the higher fares and spatial locations of the urban poor, these rapid transit systems are often difficult for the urban poor to reach and use.

This paper intends to analyze the urban transportation strategies and their impacts on the urban poor. It first discusses the definition of the urban poor in China, describes the travel characteristics of the urban poor. It then compares urban transportation development strategies in two large cities, Beijing and Shanghai, and their impacts on the urban poor. Finally, some policy recommendations are made to address the transportation needs of the urban poor.

Who are the Urban Poor in China?

The urban poverty issue in China has long been regarded as unimportant and of no great concern. Until the 1990s, poverty had traditionally been considered strictly a rural problem. Although poverty existed in urban areas, it affected only a small portion of the population and was not considered significant. Research conducted by the World Bank indicated that the urban poor accounted for less than 1 per cent of the urban population, from 1978 to 1990. Therefore, social policies dealt mainly with helping the rural poor rather than addressing the needs of the urban poor.

There were very few research reports available to describe the conditions, needs and characteristics of the urban poor in China. It was not until the 1990s the urban poverty issue started gaining the attention of Chinese governments and researchers. But there are still no clear definition and criteria offered by the government, and scholars are still debating on what constitutes the "Urban Poor".

To define the urban poor, we have to first define the urban population. Unlike other countries, the urban population in China is not defined as residents living in the urban area, but rather, it is defined as the population who are registered as urban residents (non-rural residents). There are many people who live in urban areas but are not "registered" as urban residents and therefore are not counted within the "urban population." In China, not everyone can be registered as an urban resident; registration has to be approved by the government. This is the "Hukou" management system to control or limit rural residents migrating to urban areas. On the other hand, some registered urban residents may not necessarily live in urban areas. In 1999, 389 million urban or non-rural residents were counted in the census; 40 per cent were living in the suburban or exurban areas, while 38 per cent were registered rural residents (95.6 million) living in urban areas. The difference between registered urban residents and rural residents is very important in determining the numbers of urban poor. For example, the poverty line for registered urban residents is roughly 1800 Yuan RMB ($218) a year, while for registered rural residents it is 635 Yuan RMB ($77), a difference of three times.

POVERTY LINE AND ASSISTANCE LINE

In principle, the poverty line in China is based on the minimal expenditure needed for a socially acceptable living standard. But the concept of the "socially acceptable living standard" is not accurately defined, and differs from city to city based on different levels of cost of living. In practice, the poverty line is defined as the income needed to buy a specified amount of food plus a few essential nonfood items. In the 1990s, the poverty line was calculated by a number of Chinese institutions including the National Statistical Bureau, Ministry of Civil Affairs (MOCA), and the Institute of Forecasting of the Chinese Academy of Science, using different models and criteria. The most commonly used figure for the poverty line in urban areas is between 1700 to 2400 Yuan RMB ($206-$291) per year.

The purpose of setting up the poverty line is for the government to assist those people and households whose income is below the poverty line. By delineating an assistance line, the government can determine which people and households need assistance. There is a difference between a poverty line and an assistance line. Not all people whose income is below the poverty line can receive government assistance. In 1997, the Chinese Central Government established the urban Minimum Living Standard Scheme (MLSS) to assist the urban poor.

The MLSS requires each municipality to determine an income level below which the residents should get financial assistance from the local municipal governments. This income level is called the assistance line. The assistance line is set by local municipal governments based on the local cost of living, the availability of local funds for assistance, or other factors. For example, it might mean that the assistance line is lower than the minimum wage as well as an unemployment severance fee.

Urban residents whose incomes fall below the assistance line are qualified for assistance. Because the assistance line is set by local municipal governments and the available funds for assistance from local governments differ greatly, the assistance line varies dramatically from city to city. For example, in September 2000, the assistance line in large cities like Beijing and Shanghai was between 2400 to 3828 Yuan RMB ($291-464) a year per person, but in small cities, it was less than 1320 Yuan RMB ($160). In 2002, the assistance line for Beijing was 290 Yuan RMB ($35) per month and for Shanghai was 280 Yuan RMB ($34) a month. Table shows the number of persons under the official assistance line in four major cities.

Table. The Number of Persons and Households that Received the Social Assistance Benefit by Select City in 2002.

Category/City	Beijing	Tianjin	Shanghai	Chongqing
Urban poor (total person)	119,583	301,393	431,557	718,258
-- Employed (person)	13,867	17,902	240,480	48,878
-- Laid-off (person)	7,588	74,345		71,471
-- Retired (person)	3,405	7,010		26,876
-- Unemployed (person)	10,752	82,808	170,221	116,322
-- People without proper identifications * (person)	5,763	2,481	1,724	23,603
-- Others (person)	78,208	11,6847	19,132	43,1108
Household (household)	54,491	115,221	288,678	347,857
Non-rural population (1000 person)	8,069	5,411.4	10,188.1	7,214.5
Urban poverty rate (%)	1.48%	5.57%	4.24%	9.95%

Note:

*People without personal ID card, residence card or working permit

The official number of the urban poor based on the assistance line significantly underestimates the number of persons below the poverty line. Based on a survey in the City of Bengbu conducted by the Minister of Civil Affairs of China, the actual number of people under the poverty line is twice the reported number by the local municipal government. This survey defined three major reasons for the undercount.

First, as mentioned above, the assistance line is determined by the local municipal governments based on the availability of the social assistance funds in each municipal government. There is a tendency for the local governments to arbitrarily lower the poverty line to reduce their own burden of financial assistance. Furthermore, this is compounded by the fact that the number of urban poor is based on reports from lower levels of governmental units rather than on surveys which tend to more accurately represent the number of urban poor. Many households whose income is below the local assistance line were not even aware of the MLSS program and thus did not even ask for assistance. Also, some households that applied for the MLSS program did not get any assistance because of insufficient assistance funds in the local governments. Second, the official number of the urban poor does not include rural residents who live in urban areas (the so-called mobile residents). Most of these rural residents who live in urban areas have even less income than the qualified urban poor. Based on an analysis by the Asian Bank, the poverty rate within the mobile population is about 50-100 per cent higher than the permanent urban residents.

Third, the urban poor are determined on the basis of whether or not income is below the assistance line. This is misleading because the expenditure required to maintain a minimal living standard or the socially acceptable living standard may be greater than the poverty line determined by income. The urban poor may cope by either lowering the minimal living standard, or by dipping into savings. According to a study by the Asian Bank, if the expenditure of acceptable living standard is used to measure the poverty rate, the number of persons who qualify as the urban poor would be 2.5 times higher than the reported rate. In Beijing, that would be about 7.7 times more. Taking into consideration both the expenditure and number of rural residents in the urban areas, the poverty rate would be 11.6 per cent in Beijing and 6 per cent in Shanghai, which are much higher than the rates shown in Table.

Main Causes of Urban Poor

Who are these urban poor and what causes these people to become the urban poor? For the most part, they are laid-off workers, the unemployed, the retired, the working poor, people with disabilities or chronic diseases, and a portion of the rural migrants. The urban poor in China are generally the result of growth in unemployment, "forced" early retirement, as well as the influx of farmers to the urban areas. In a nutshell, lack of jobs is the major cause of the urban poor. Since 1994, a large number of governmental workers have been laid off due to the reorganization of government-owned businesses. For example, between 1996 and 1999, employment in the government-owned businesses has decreased by 27 million jobs, or 25 per cent of the total employment in government-owned businesses. At the same time, job growth in the private sector was unable to absorb these work forces.

In 2000, the official unemployment rate in urban areas was 3.1 per cent. But this number only includes registered unemployment, and many others are not included in the estimate. Of particular importance are workers that were laid off. When workers are laid-off, they are usually given a very small amount of living expense, but they are not counted as "unemployed." In 2000, there were 8.6 million laid-off workers in China. In addition, there were other workers at smaller factories forced into early retirement. These people were also not counted in the official unemployment rate. Adding together these laid-off workers, the registered unemployed, and other unemployed, the actual unemployment rate could be as high as 11.5-15 per cent. This great number of unemployed becomes a major source of the urban poor. They have unique travel needs which should be addressed in any urban transportation plans and development strategies.

URBAN POOR TRAVEL AND HOW MUCH DO THEY SPEND IN TRAVELING

How do the urban poor travel? Unlike the United States, transportation survey data in China usually do not include questions on socio-economic variables. It is thus difficult to analyze the travel characteristics of the urban poor based on available transportation survey data. However, this information can be derived from other data sources, as well as field observations. Table shows the travel expenditure of low income and very low income residents, who may not necessarily qualify as the true urban poor. Households with low income and lowest income spend less on transportation, compared with the average and other income groups.

This is because most low-income residents rely on walking and biking as major travel modes. In fact, the expense of transportation fare by low and very low income households is about 11 Yuan ($1.33) in Beijing and 15 Yuan ($1.82) in Shanghai per month. Bus fare costs one Yuan RMB ($0.12) a ride. What does this mean? If the average person takes two bus trips per day, and if each trip requires 1.88 transfers on average, and each transfer costs an additional one Yuan RMB, that person needs to spend 3.76 Yuan RMB ($0.46) a day. The 11 Yuan RMB can only be used to take buses for three days in Beijing. This is a strong indication that the urban poor rely on non-motorized modes in their daily life.

With an average income of 280-290 Yuan RMB per person per month ($34 -$35), the urban poor cannot easily afford bus fare. Based on the above calculation, the total transportation cost is 3.76 Yuan RMB per day. That would be about 112.8 Yuan RMB ($13.67) a month, which is about 40 per cent of the monthly income. This is in fact a conservative number since the average number of daily trips taken in Beijing is about 3.59 per person. This is obviously not affordable.

In fact, the average proportion of the transportation expenditure over discretionary income in Beijing, Shanghai, and the nation as whole has increased steadily over the last few years. But the rate of increase for low-income households is even faster than that of the average households in both Beijing and Shanghai.

The urban poor will continue to spend an increasingly greater portion of their monthly income on travel. This trend will continue in the future, which will put even greater pressure on the urban poor. At the same time, free or less expensive modes of transportation such as walking and bicycle riding are only suitable for short distance trips, and only when pedestrian and bike paths are available.

As cities expand to the fringes, housing and employment opportunities grow further apart. Walking and biking as travel modes become less viable. If biking and walking are not feasible, either due to transportation policy or due to long commuting distance, transportation costs would be much higher. That is the case in the United States.

Table. Annual Expenditure of Urban Residents on Transportation, 2002

Index		All Cities		Beijing		Shanghai		
	Average	lowest income	Low Income	Average	Low Income	Average	Lowest income	Low income
Households Surveyed (Household)	45317	4532	4532	1000	100	500	50	50
Discretionary Income (Yuan RMB)	7702.8	2408.6	3649.16	12,463.9	6,057.5	13,250	5,791	7,574
Total transportation expenditure(Yuan)	267.24	52.94	92.98	684.5	490.8	555	213	282
--Transport Fare (Yuan)	146.04	34.25	56.68	300.1	138.3	388	181	223
--fixed costs of purchasing vehicles and/or bicycles (Yuan)	78.96	10.51	19.53	291.2	296			
--others (Yuan)	42.24	8.18	16.77	93.2	56.5	167	32	59
Proportion of the expenditure on transportation to the discretion Income	4.43%	2.22%	2.85%	5.49%	8.10%	4.19%	3.68%	3.72%

Note: The lowest and low income is the lowest 10 and 10-20 per cent of the income (Beijing combines the two categories into one).

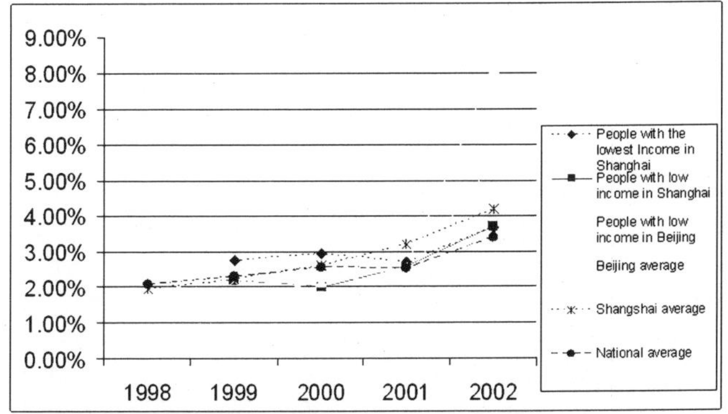

Fig. The Shift of the Ratio of the Expenditure on Transportation to Discretionary Income Per Capita

In the United States, the working poor are often defined as living below the poverty line ($8500 in 1999 according to the US Census Bureau, or $8000 according to the Bureau of Transportation Statistics). The working poor use less expensive commuting modes such as public transit, walking, and biking more often, and spend much less to commute than higher income workers who typically own more than one vehicle per household. However, in the United States, the working poor spend a significantly greater portion of their income on commuting and travel than any other income group. As income levels rise, the portion of income spent on travel significantly decreases. The working poor spend approximately 9.5 per cent of their income on commuting. Although considerably greater than most other groups by income in the United States, this is still a significantly small income proportion in comparison to that spent by the urban poor in Beijing or Shanghai.

Table. Median Per cent of Personal Income Spent on Commuting by Income Group in the United States, 1999.

Annual Income	% Spent on Commuting
Less than $8,000	9.50%
$8,000 to $14,999	6.00%
$15,000 to $21,999	4.60%
$22,000 to $29,999	4.10%
$30,000 to $44,999	3.50%
$45,000 and Higher	2.20%
Total Population	3.90%

Table. Median Per cent of Personal Income Spent on Commuting in the United States by Income Group and Mode, 1999.

Annual Income	Own Vehicle	Public Transit
Less than $8,000	20.50%	12.80%
$8,000 to $14,999	8.00%	5.70%
$15,000 to $21,999	5.60%	3.90%

$22,000 to $29,999	5.10%	3.00%
$30,000 to $44,999	4.20%	2.20%
$45,000 and Higher	2.60%	1.50%
Total Population	4.90%	3.30%

In summary, the urban poor currently rely heavily on walking and biking as vital transportation means. But as urban areas expand and commuting distances increase, walking and biking will become more difficult. Some urban poor will be forced to take bus or rail, which will put great pressure on poor people's discretionary income. The urban transportation policy should carefully consider the accessibility needs and the affordability of the urban poor. But how do the policy-makers consider the interests and needs of the urban poor?

URBAN TRANSPORTATION STRATEGIES AND THEIR IMPACTS ON THE URBAN POOR

Congestion relief and travel time reduction are the key problems facing Chinese cities. In response to these problems, government policies are focused on creating modernized systems to reduce congestion and travel time. But in the decision-making process, not enough consideration is given to the needs of the poor. This section discusses urban transportation strategies in large Chinese cities, particularly Beijing and Shanghai, and their impacts on the urban poor. It starts with a description of the general trends of travel modes in those cities, particularly the trends of walking, biking, and the use of transit. It then discusses the urban transportation strategies by mode, from automobile, to transit, and to non-motorized modes.

RECENT TRENDS OF URBAN TRAVEL IN CHINESE CITIES

Data in the past two decades show that walking and biking are overwhelmingly important means of travel for many urban residents, but their shares are starting to decline. Public transportation is the next most important travel mode, and its share has remained steady. Private cars are starting to gain importance as a viable commuting mode as ownership increases.

Table. The Modal Split in Selected Large Chinese Cities (percentage)

City	Year	Walk	Bicycle	Public Transit	Commuter bus of employers	Private automobile	Taxi	Motor cycle	Other
Beijing	2000		38	27		23	9		3
	1986		58	32		5	1		4
Shanghai	1999	30.74	39.01	15.16		15.09	NA	NA	NA
	1995	30.11	41.18	17.42		11.29	NA	NA	NA
	1986	41	31	24		4			
Guangzhou	1998	41.92	21.47	17.09	1.41			0.72	10.35
	1984	39.1	34.05	19.37	9.61	0.87	0.27	0.37	

Nanjing	2002	23.23	43.79	24.74	0.95	0.42	1.01	2.85	3.12
	2001	26.5	41	24.4	3.1		1	2.7	1.3
	1999	23.57	40.95	20.95	5.68		1.71	5.24	1.89
	1997	25.45	57.91	8.19	4.51		0.92	2.16	0.68
	1986	33.1	44.1	19.2	2.5		0.1	0.3	0.7

Walking and biking have been the primary means of transportation but their shares have decreased over the years. For example, in 1986, biking accounted for 58 per cent of the share in Beijing without accounting for short distance walking trips. In 2000, walking accounted for 33 per cent, biking 39 per cent, and public transit only 14.4 per cent of the total share in Beijing. Compared to the 1986 data, if short walking trips are not taken into account, biking would account for 38.49 per cent, and transit 26.51 per cent. In Shanghai, walking and biking account for 72 per cent in 1986, but it dropped to 69.75 per cent in 1999. Similarly, in Nanjing, the walking and biking share has dropped from 77.2 per cent in 1986, to 67.02 per cent in 2003, and in Guangzhou, the share has dropped from 73 per cent in 1984 to 63.4 per cent in 1998. Another noticeable change shown in Table is the decreased share of public transportation. In Beijing, the share of public transportation (including bus, rail, and minibus) decreased from 32 per cent in 1986 to 27 per cent in 2000 after adjusting for short distance walking trips, despite the investment and expansion of the metro rail system.

There are several reasons for the declining role of the walking and biking share. First, as the city continues to expand outward and travel distances increase, walking and biking become more difficult. Second, some walkers and bikers were lured away by improved transit services in some cities. For example, from 1997 to 1999 in Nanjing, the public transit was improved and the local roads were not very congested. Buses could maintain a reasonable speed and hence attracted many bicycle riders. But overall in China, transit share did not increase, and the increase of transit share in Nanjing may be temporary. In recent years, the roads have become more congested, inhibiting the speed of buses. Many people still find biking is a faster transportation mode to get around.

In Nanjing, the most remarkable change of the modal split is that residents began to use more public transit after the mid 1990s. The percentage of trips made by public transport increased from 8.19 per cent in 1997 to 24.74 per cent in 2002. Conversely, the walking share fell from 33.1 per cent to 23.23 per cent. On the other hand, the trips made by bicycle maintained the greatest share (more than 40 per cent) among all modes during the past 17 years. Another change occurred in 2002; the private auto was listed as a separate transportation mode for the first time and was no longer aggregated as one of the other modes. In the background of the mode shift over the years in big cities in China, particularly the dramatic increase of car ownership and automobile usage, and corresponding roadway congestion, many cities have developed future transportation strategies that are based on projected future needs. The transportation strategies categorized by modes will be discussed, their impacts on the urban poor will be analyzed, and some countermeasures to serve the interests of the urban poor will be proposed.

THE DEVELOPMENT OF THE EXPRESSWAY NETWORK AND ITS IMPACTS ON THE URBAN POOR

Rapid Growth of the Automobile

The automobile industry has been identified as the leading industry for economic development and employment growth in China. The Central Chinese government encourages automobile ownership and usage by providing car loans. Most municipal governments also encourage automobile ownership. The City of Shanghai is the only one that places limits on the numbers of licenses granted on new cars. This system is similar to one in Singapore. In addition, the media, the car manufacturers, and the people all consider car ownership the premier symbol of social status. Owning a car has become a dream of most people. With economic growth, the income of some urban residents has also increased, and as a result, private auto ownership has skyrocketed over the last few years. For example, Beijing's private car has increased from 173,600 in 1996, to 1,280,000 in 2003, an increase of over 7 times.

Table. Private Automobile Ownership from 1996 to 2003 in China, Beijing, and Shanghai.

City	Year	1996	1997	1998	1999	2000	2001	2002	2003
Nation	Total auto (10000veh)	1100.08	1219.1	1319.30	1452.94	1608.91	1802.04	2053.16	
	Private auto (10000veh)	289.67	358.36	423.65	533.88	625.33	770.78	968.98	
Beijing	Total auto (10000veh)	62.18	78.43	89.85	95.14	104.11	114.47	133.93	202
	Private auto (10000veh)	17.36	29.76	40.75	44.74	49.41	62.40	81.08	128
Shanghai	Total auto (10000veh)	34.28	38.34	38.68	42.54	49.19	55.0	62.3	
	Private auto (10000veh)	0.92	1.000	0.92	2.46	5.07	8.72	14.68	22.3
	Private auto (10000veh)	0.77	1.08	1.44	1.91	2.74	3.76	5.14	8.3

The dramatic increase of automobile ownership has placed increasing pressure on urban road systems. Most of the systems were designed before the automobile age and thus are ill prepared and inadequate to meet the demand. As a result, traffic congestion and gridlock have become commonplace. To combat the traffic problem, many cities have decided to expand the road system by developing expressways and major arterial roads, including elevated roadways.

For example, from 1996 to 2003, the total expenditure in urban transportation infrastructure in Beijing was 116.1 billion Yuan RMB, approximately 5.88 per cent of the city's GDP. As a result of this vast expenditure, the urban transportation infrastructure has improved significantly:

- Urban roads increased from 3665 km in 1996 to 3786 km in 2003, a 3.3 per cent increase in length; the area of roads increase from 38.07 million square meters to 61.5 million square meters, a 61.5 per cent increase over same time period. This indicates that the major road expansion in urban areas (central city) involves widening existing roadway.
- Roadway in the Beijing metropolitan area experienced a 23.7 per cent increase, from 11,682 km in 1996 to 14,452 km in 2003. The urban expressway increased by 339.5 per cent, from 114 km to 501 km during the same time period. This shows the bias in transportation infrastructure in the development of urban expressways.

In Shanghai, the expenditure in urban infrastructure increased, on average, approximately 40 per cent each year, from 878 million Yuan RMB in 1991, to 9.6 billion Yuan RMB in 1998. As a result, the urban road increased from 4818 km in 1991 to 6678 km in 1998, a 40 per cent increase. The total road area in 1998 was 9763 square km, which is about 2.6 times of that in 1991.

In the future, this road building frenzy in both Beijing and Shanghai will exacerbate, especially stimulated by the Olympics in Beijing in 2008 and the Worlds Fair in Shanghai in 2010. In Beijing, the overall transportation development strategy emphasizes the development of new expressways, particularly in the suburban and exurban areas. The plans include building 390 km of new expressways and 1000 km of new arterial roads, at the cost of about 30 billion Yuan RMB. Inside the urban area, the plan is to build an additional 65 km of expressway and more than 200 km of urban arterial roads. In addition, more plans have been made to build and improve large interchanges.

This emphasis on urban expressway and arterial roads will do little to improve the accessibility of the urban poor. In fact, the large scale of road building limits resources to public transportation and non-motorized modes. Since only a very small portion of households are able to afford a car, this biased expenditure in highways and expressways raises the issue of equity: the money used to build more roads does not come from a user fee, but rather, it comes directly from the general municipal funds. Current road building expenditures disproportionately benefit the few auto users, at the expense of non-auto users. Furthermore, the effectiveness of building new roads in order to "build their way out of congestion" is questionable given the experience in other countries of the world.

Strategies of the Public Transit and their Impacts on the Poor

Public transit has been considered a priority in Beijing and Shanghai. The major strategy however, is to develop rapid transit systems like metro rail, light rail, or rapid bus transit. Although seen as important, the bus system is not considered as high a priority as the urban rail system. As it shares the right-of-way with automobiles, bikers, and pedestrians, the importance of the bus system has fallen.

Rapid Transit Systems

Rapid transit systems include metro rail, light rail, bus rapid transit, and the maglev systems. These rapid transit methods are given a much higher priority and are more favored than the bus system because they are seen as symbols of a modern, technologically advanced society. They are also considered to be tangible achievements of politicians. In the 1990s, urban rail systems in several large Chinese cities developed very rapidly. For example, the metro rail system in Beijing increased from 41.6 km in 1996 to 114 km in 2003, a 174 per cent increase. In Shanghai, the urban rail system was initiated in the early 1990s, and as of 1998, included 78.4 km of line.

In the next few years before the Olympic Games, the City of Beijing plans to build about 186 km of urban rail line, and 60 km of rapid bus transit system (RBT). The expenditure in these rapid transit systems will be huge. For example, in 2003, the expenditure in the rail system was 8.45 billion Yuan RMB, or 40.7 per cent of all urban transportation infrastructure expenditure. It sets up a very ambitious goal of raising the transit share from the current 27 per cent to 60 per cent, with the rail and BRT system accounting for 40 per cent of transit ridership. Spending on the rail system in Shanghai is more significant. In 2002, it accounted for 44.8 per cent of all urban transportation infrastructure expenditure (14.29 billion Yuan RMB). In 2002, the spending on rail systems was 12.47 billion Yuan RMB (including 5.58 billion Yuan RMB in the Maglev system; the total cost of the Maglev was about 10 billion Yuan RMB), or 56.8 per cent of all transportation expenditure in Shanghai (excluding Maglev, 31.4 per cent). In the future, Shanghai will invest around 2.5 per cent of its GDP to construct more than 500 km of rail line before 2020.

The love of rail system has spread to other cities. For example, the City of Nanjing will build a 120 km rail transportation network consisting of subway and light rail within the next 15 years. Based on the experience of Japan, the development of urban rail and rapid bus transit makes sense in a dense urban area with a monocentric land use pattern, as in the case in Beijing and Shanghai. But the benefits to the urban poor are very limited in the short run, and are not clear in the long run because the urban poor simply cannot afford it unless the government provides more transportation assistance. In Beijing, the one-way metro rail fare without a transfer costs 3 Yuan RMB. With a transfer, it costs 5 Yuan RMB. Light rail fare is 3 Yuan RMB. A transfer from the metro rail to light rail incurs a separate fare (*i.e.*, 3 Yuan each). As mentioned before, based on the survey data in Beijing, the average transfer rate is 1.88, the two-way rail trip will cost about 2*5*1.88/2 = 9.40 to 2*6*1.88/2 = 11.28 Yuan RMB. The monthly cost would be between 282 to 338 Yuan RMB, which is close to or exceeds the assistance line of 290 Yuan RMB per month.

Similarly in Shanghai, the fare of the metro rail is based on ride distance; the minimal cost is 2 Yuan RMB for the first 6 km, 3 Yuan for 16 km, after that every 6 km costs an additional 1 Yuan, with the maximum of 6 Yuan for a one way trip. For the light rail, the fare is based on a zonal system with 2 Yuan and 3 Yuan zones. This is significantly more costly than the flat 1 Yuan bus fare. If the bus fare is too high for the urban poor to afford, the rail fare is far beyond the reach of the urban poor.

An extreme example of striving for modernization, speed, and tangible achievement is the development of the world's first commercial magnetic-levitation train. The 30 km long Maglev in Shanghai was first opened to the public on September 30, 2003 and started regular operation in 2004. The Maglev uses a powerful magnetic field to suspend trains millimeters above their rails and propel them at a speed of 430km/h. However, there are a number of technical and operational problems in the design and building of the Maglev system in Shanghai.

First, portions of the track were found sinking in mid-April of 2004, which will require more capital expenditures. Second, the Maglev has also suffered from low ridership. Ticket sales have been very slow since its opening. During the first week of regular operation, each train only carried, on average, 73 passengers, much below its capacity of 440. The overall occupancy rate is only about 10 per cent. This led to a fare reduction on April 15th, from 75 Yuan RMB to 50 Yuan RMB one way, and 80 Yuan RMB two way. But even with the price cut, the ridership increase has been marginal. It is expected that the fare will continue to drop in the near future in order to attract more riders. The third problem involves the bad planning of connections with other transit systems. The Maglev is not well-integrated with Shanghai's other transit systems. The start station is located miles away from the heart of downtown and has poor transfer facilities. All these problems lead to the unsuccessful operation of the Maglev. The high price of the system almost guarantees the exclusion of the urban poor, as well as most middle-income passengers, to use it as a means of daily travel.

Bus System

The bus system has played a moderate role in the urban transportation system despite the spending increase over the years, but its importance may decline in the future. For example, in Beijing, the number of bus lines increased by 42 per cent, from 383 routes in 1996 to 544 in 2003. The number of buses increased from 4781 vehicles in 1996 to 16,634 vehicles in 2003, an increase of 247.9 per cent. But at the same time, transit ridership (including rail) only increased 26.9 per cent from 3.5 million to 4.44 million. The future of the bus system is not very optimistic. Some big cities like Shanghai plan to reduce the route length of the bus system from 23,259 km to around 12,500 km, a 46.26 per cent reduction. Many bus routes will also be eliminated, and the system may continue to reduce its share of travel.

There are three major reasons. First, due to the ever increasing road congestion, the travel speed of buses has steadily declined. With an average speed of 9 to 10 km per hour, a short-distance rider will shift to bicycles and motorcycles. A bike would be faster and more flexible than the bus. For a long-distance rider, the rail would be a much better option, as it can have significant time savings over the bus. Third, the bus system is a boring and unexciting travel mode for politicians to show off the modern image of their cities and their tangible achievement. Politicians are less likely to have strong incentive to promote or endorse the development of new bus systems in cities. This bias against the bus system will have very negative impacts on the urban poor. For them, the bus system is the next best mode to biking for commuting and mobility. Failure to improve the bus system will put the urban poor at an even greater disadvantage.

The Minibus System

The minibus system is similar to a jitney or paratransit system. It is usually operated by private individuals, using small buses or minivans. It is considered an effective supplement to the public transit system, serving mainly the remote suburban areas. It is flexible and can stop at any place along a more or less fixed route to pick up or drop off passengers. The minibus system has been serving cities in China well since the 1980s when the system was introduced.

But its fate is different in different cities. In many big Chinese cities, such as Beijing, Nanjing, Changsha, and Xi'an, minibuses are stigmatized as an unsophisticated mode of transport, which is not only in conflict with the public transport priority strategies, but also harmful to the images and reputations of the cities. Therefore, minibuses are only allowed to run in suburban areas in Beijing, Nanjing, Changsha, and Xi'An now. In other cities like Shenzheng, Chongqin, the minibus is still used as an important part of the urban paratransit system, while Shanghai and Guangzhou do not have minibus services.

In the future, the minibus will still have its place in the overall public transportation system, especially in the environment of increased low-density development in the suburbs. According to the transportation strategy of Beijing, minibuses are considered a part of the public transit system, but will only be placed in areas where rail transport cannot reach. But in the city of Nanjing, minibuses will continue to be used on all local streets to serve the urban community. If managed well, minibuses may become an effective supplemental mode for buses and rails.

The fare of the minibus is higher than the regular bus, but it provides service to areas that regular buses do not, at a reasonable price. The pricing structure of minibus service is regulated by the local municipal government. For example, in Beijing, the starting price for a minibus ride within 5 km is 2 Yuan, with an additional 1 Yuan increment for every 4 km. The maximum fare for any ride is 6 Yuan RMB. Some minibus services provide long-distance transport up to 80 km. The 6 Yuan maximum charge is even cheaper than the regular long distance bus fare. This is a service that the urban poor depend on for traveling to areas without regular bus service, particularly to long-distance remote areas.

In summary, development strategies in urban public transportation in big Chinese cities like Beijing and Shanghai focus mainly on the development of rapid transit systems like metro rail, light rail and bus rapid transit systems, while the regular bus system is fated to decline. These strategies will not benefit the urban poor due to the extremely low income of the urban poor unless the government provides more transportation assistance to allow the poor to use these rapid public transit services free of charge or with reduced fare. The minibus provides a necessary supplement to the regular bus services, but measures should be taken to broaden the coverage area.

Strategies of Biking and Walking and Impacts on the Poor

Despite the fact that the vast amount of urban commuters depend on biking and walking, their roles in the overall urban strategies are not given enough attention in most, if not all, urban transportation development strategies. These two transportation modes have been considered secondary and unexciting, and some may even consider them to be backward and incompatible with the image of progressive modernization. But to the urban poor, biking and walking have been and will continue to be important means of their daily trips. It is generally believed that biking is the most preferred method of transportation for short distance trips, especially in congested urban areas. It is clean, convenient, economical, flexible, and healthy. It contributes to about 30 per cent to 50 per cent of all trips in large Chinese cities and only occupies a small part of the road system. Bicycle ownership is almost universal for residents in Chinese cities, even for poor families. In 2003, it was reported that the average number of bicycles owned by poor households is 2.3 in Beijing, which is even higher than the city average. When public transit becomes unaffordable for the urban poor, bicycles are the most important means of travel.

Although bicycle riding has many advantages, it is slow and has caused some problems for roadway traffic. Large volumes of bicycle traffic have caused serious problems at street intersections, and bicycles have become a significant source of severe traffic jams. In addition, some planners and governmental officials consider such widespread use of bicycles as incompatible with the ideal image of a modern city. Therefore, some cities started to restrict or even prohibit bicycles on major city roads or in the central city. In Nanjing, Guangzhou, and some other cities, the previously planned bicycle-only lanes and streets were canceled and bikes are currently restricted to sidewalks. As a result, bike riders have to search for new routes and travel longer distances to get to their destinations. This is consistent with many transportation strategies. For example, the cities of Shanghai and Nanjing have set up goals to cut the share of bicycle trips by half to about 20 per cent~25 per cent of total trips.

In spite of that, biking continues to be a major transportation mode in Chinese cities, because it is cheap, flexible, and even faster than buses and cars in very congested central cities, particularly during rush hours. For example, in 2001, one year after the rules of restricting bicycle usage on many city roads were enforced, the bicycle trip share increased, and was even higher than in 1999. The data in 2002 also did not show a falling trend of biking in mode share.

There are, however, some positive signs indicating that some cities are trying to make bicycling a viable transportation mode. For example, in its 2000 transportation strategic plan, Shanghai proposed to build more bike lanes on local streets and on bus-only streets. In addition, more secure bicycle parking facilities will be provided in the city, and more park-and-ride facilities for bikers will be built near public transit stations. Similar planning is also being implemented in Beijing, Nanjing and other big cities. The construction of bike lanes and bike-only streets, secure bike parking and park-and-ride facilities would facilitate the use of bicycles in the city, which will help the urban poor who rely disproportionately on non-automobile travel modes. But overall, bicycle riding is confined and discouraged. Just like cycling, pedestrian facilities are lacking in most Chinese cities, and walkers have trouble crossing roads and intersections. In most big cities, high numbers of parked bicycles often encroach on footpaths. In some roads, pedestrians even have to share the right-of-way with bicycle riders. The walking environment has been neglected and has never been fully developed. It has even worsened in recent years despite public calls for the protection of pedestrians. The walking environment is so dangerous to pedestrians that walking across the street can be highly risky.

Table. Traffic accidents in Shanghai in 2002.

	Total	Motor Vehicles			None-motor modes			Pedestrians and Passengers*
		Automobiles	Motor cycles	Tractors	Others	Total	#Bicycle	
Number(case)	46733	39780	2138	97	221	3029	2334	1378
Death(person)	1398	616	204	3	14	351	295	210
Injuries(person)	15585	9189	2265	48	71	2765	2131	1247
Losses Converted into cash (10000Yuan)	29754	28874.30	365.1	56.04	117.07	262.73	200.98	78.95

Note:

*This number includes both pedestrians and bus passengers.

In fact, there is a high rate of crashes involving bicycles and pedestrians as shown in Tables 7 and 8. For example, in Shanghai, of the 1378 deaths caused by traffic crashes, about 36 per cent (505 cases) involved bikers and pedestrians (including bus passengers) in 2002. Similarly, in Beijing, 27.7 per cent of deaths were bikers and pedestrians in 2002.

Table. Traffic accidents in Beijing in 2002.

	Total	Motor Vehicles	None-motor modes
Number(case)	17645	15190	2455
Injuries (person)	10424	8226	2198
Death (person)	1447	1046	401

In addition, pedestrian planning in Chinese cities remains focused on the streets in the downtown, commercial districts or in scenic spots, places where walking is more likely to be viewed as a recreational pursuit rather than a mode of transport.

Although some measures, such as constructing road crossing facilities and setting pedestrian facilities in the communities, were mentioned in the transportation strategies of Shanghai, little attention was paid to the 'pedestrianization' of urban life. Since walking is not encouraged in big cities in China, and it is less convenient when travel and commuting distances are both increased, its trip share is projected to decrease significantly in the future.

Many transportation planners and engineers perceive bikers and pedestrians as major contributors to traffic congestion. Often, much of the congestion is caused when rules are disobeyed. There are many issues concerning the enforcement of traffic rules, especially for bikers and pedestrians who traditionally do not follow the rules rigidly. But this is not a reason to abandon or restrict bicycle use. On the contrary, transportation planners, engineers and policemen should work together to find traffic control solutions and rules that serve bikers and pedestrians better.

IMPACTS OF SUBURBANIZATION ON THE URBAN POOR

Suburbanization has had a serious impact on the mobility of the urban poor in China. The transition to a socialist market economy in the late 1970s caused major population increases as the rural poor flocked to the cities. The 'commodification' of land greatly influenced municipal government development policies and has affected the urban poor in two significant ways. First, there has been great pressure to redevelop urban core areas, to generate greater revenue by replacing urban poor residences with commercial or high-end residential buildings. This has had a significant impact on existing urban residents who no longer can afford to stay in the core. As an added incentive, the government also provides subsidies to the urban poor to help them move to the suburbs or exurban areas. But many of the urban poor have retained jobs within the core, increasing their commuting times, as well as congestion.

Table. Population and Number of Employed Persons by Region in Beijing (in millions)

	Total Population		Central Areas		Near Suburbs		Exurban Areas		Rural Counties	
Year	Total Population	Total Employment	Core Population	Core Employment	Near Suburban Population	Near Suburban Employment	Exurban Population	Exurban Employment	Rural Population	Rural Employment
1991	11.157	4.6854	2.59	1.5882	4.081	2.2317	1.041	0.2715	3.445	0.5940
1996	11.840	4.4256	2.671	1.4141	4.573	2.1831	1.030	0.2311	3.566	0.5974
1997	12.167	4.3738	2.691	1.3649	4.821	2.2027	1.051	0.2230	3.604	0.5832
1998	12.234	4.4988	2.651	1.4404	4.933	2.2895	1.665	0.2904	2.985	0.4785
1999	12.499	4.3777	2.654	1.4384	5.146	2.1815	2.727	0.4825	1.972	0.2725
2000	12.780	4.2858	2.663	1.3396	5.373	2.2223	2.778	0.3937	1.966	0.2767
2001	13.666	4.3605	2.793	1.3137	5.896	2.2598	3.580	0.5932	1.397	0.1653
2002	14.952	4.7761	2.859	1.1663	6.589	2.4950	4.778	0.9116	0.726	0.1303

Note:

1. The population here includes the permanent residents and the temporary residents.
2. Rural population here refers to the population living in rural counties.

Throughout the 1990s in Beijing, the overall population grew from 11.16 million to 14.95 million. Although the urban core population increased from 2.6 to 2.9 million, it was marginal compared to the growth in the near suburban and exurban areas and which experienced a 38 per cent and 78 per cent increase in population respectively. In 1999, the exurban population of Beijing surpassed the urban core population, a strong indication of suburbanization. In comparison, between 1991 and 2002, total employment fluctuated, and only increased marginally throughout the entire Beijing area from 4.7 to 4.8 million, an increase of only 100,000 jobs, an indication of increasing unemployment, a major source of urban poverty. The second major impact of suburbanization involves county governments in the periphery of large cities encouraging development in peripheral areas where land prices are inexpensive in order to increase revenue. Most of the development in the periphery has been residential, and much of it has been for the poor. The incoming rural poor are at an even greater disadvantage than the native urban poor, and are often forced to stay in peripheral slums. Combined, these agendas promoted hasty outward expansion while the government failed to address issues concerning the great increases in commuting and congestion.

Table also shows the distribution of employment by area in Beijing. Employment growth in all areas is slower than the population growth, especially in the urban core. The employment at the core area decreased from 1991 to 2002. Many manufacturing jobs have since moved out to suburbs, creating problems for low-income residents who still reside in the inner city. Unlike the inner city, employment in the suburban areas actually grew from 1991 to 2002. But the employment growth rate is far below the population growth rate.

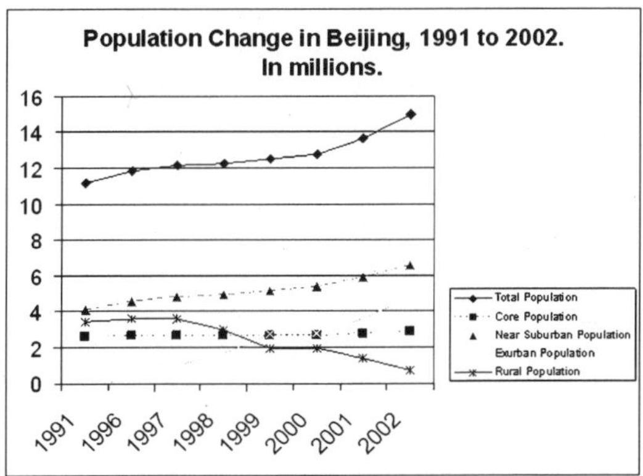

Fig. Population Change in Beijing from 1991 to 2002, by location.

A major outcome of suburbanization is that as the city size increases, the travel distance and time will inevitably increase. This causes the use of non-motorized travel mode more difficult, or even impossible, which will only make it more difficult for the urban poor to access jobs and other services.

URBAN TRANSPORT DEVELOPMENT

SIGNIFICANT IS THE URBAN TRANSPORT PROBLEM IN INDIA

Efficient and reliable urban transport systems are crucial for India to sustain a high growth rate and alleviate poverty. Indeed, the significance of urban transport in India stems from the role that it plays in the reduction of poverty, both through its indirect effects as a stimulator of poverty reducing growth and through its direct effects on the quality of life of people. Services and manufacturing industries particularly concentrate around major urban areas, and require efficient and reliable urban transport systems to move workers and connect production facilities to the logistics chain. In China for instance, service industries as well as manufacturing and labor-intensive industries have developed in economic centers endowed with good transport systems that could efficiently handle mobility needs of millions of workers and facilitate the movement of goods. India's economy is currently more service-oriented than China, with only about one million people employed in the IT industry—the mainstay of the tertiary sector. But growth in the services sector and development of Indian manufacturing industries will put more pressure on already saturated urban transport systems.

Many Indian cities such as Bangalore and Chennai have attracted significant investments in high-technology industries thanks to a competitive and highly qualified workforce. In the past few years however, urban infrastructure, and transport systems in particular, have been struggling to keep up with the growing number of firms moving into these cities. Local and international media have been continuously reporting the cities' difficulty in coping with growing demand for efficient transport systems. The financial press has been describing Bangalore for example as a city of 60,000 unfilled potholes and where software workers' morning commute to work can take up to two hours. Developing an efficient urban transport system should be part of the broad Government policy aimed at improving the attractiveness and competitiveness of Indian cities.

Responding to the needs of the economy is not just a matter of the cost of transport, but also of the quality of service provided. A 2003 Confederation of Indian Industry survey of urban populations in Southern India showed 90 per cent dissatisfied with roads, and 58 per cent dissatisfied with public transport services. Interestingly, the same survey showed that 89 per cent were willing to pay for good-quality toll roads and 65 per cent are willing to pay higher public transport fares to get more comfort and frequency. A survey of the business community recorded similar answers. Developing a degree of flexibility in public transport supply so that differentiated services may be tailored to the needs of different groups of the population is thus an important requirement of a future urban transport policy.

The impacts of transport on the quality of urban life go even further than that. In the 1990s, as India experienced a period of economic and urban growth, air pollution in its major cities became a cause of national concern and generated worldwide attention. The levels of airborne suspended particulate matter recorded in largest metro-cities far exceeded the ambient air quality standards adopted by India and many other countries. As manufacturing and power sectors are progressively cleaned up the relative importance of the urban transport sector to air pollution increases. There is much current discussion about the development of Mumbai as a "world class city" rivaling Shanghai. For the Indian cities to retain their attractiveness to international capital, and to compete with other international centers, they must be livable. The environment is important to the economic health of the cities as well as the medical health.

In parallel with the growth related impacts of urban transport on poverty are the direct impacts of urban transport on the life of the poor. The worst off in urban transport may be the pedestrians, whose mobility and safety are hindered by non-existent, broken-down, and/or obstructed sidewalks; difficult street crossings; and flooding in monsoon seasons. The bicycle riders, once a major urban transport mode in India, are gradually being pushed off busy roads by motor vehicles. These two groups account for half of all traffic fatalities. Secondary and tertiary road networks appear to have received little attention or funding, especially in low-income areas

GOVERNMENT OF INDIA POLICY RESPONSE TO THE URBAN TRANSPORT PROBLEM

The Government of India is addressing these issues. The Ministry of Urban Development, has recently issued a draft National Urban Transport Policy for consultation which can be found on their web site. They recognize the increasing urban road congestion and its associated air pollution. Their strategy puts primary emphasis on the need to increase the efficiency of use of road space by favoring public transport and by the use of traffic management instruments to improve traffic performance and by restraining the growth of private vehicular traffic. Complementing this is a strategy to reduce vehicle emissions by technological improvements in vehicles and fuels.

The Government of India's proposed strategy is in many respects along the lines of international thinking on the approach to the urban transport problem. For example, the World Bank completed an Urban Transport Strategy review, Cities on The Move, in 2002, after consultation with major stakeholders in client countries, including governments, transport operators, and nongovernmental organizations, as well as with representatives of other international institutions. That review linked urban development and transport sector strategies with a strong poverty focus. It noted that sprawling cities are making the journey to work excessively long and costly and that throughout the per capita motor vehicle ownership continues to grow with adverse impacts on traffic congestion and air pollution. Public transport is being stifled by this congestion and its relative performance tends to decline in comparison with the private modes. So the vicious circle of congestion and the decline of public transport is perpetuated. The safety and security of urban travellers are also emerging problems worldwide.

The key policy recommendations of the Cities on The Move review are not dissimilar to those of the Government of India policy paper. The review emphasizes better maintenance of road facilities and improved traffic management, and gives pride of place to the importance of public transport in addressing the burgeoning demand for movement in urban areas. Improved public transport and allocation of road space has been shown to also lead to an improved level of service for those in private vehicles (examples of Santiago and London). This requires improvement in the efficiency of operations (to be achieved through progressive involvement of the private sector in supply under strictly regulated contracts with the public authorities), priority use of existing infrastructure (including fully segregated bus rapid transit systems as well as more modest bus priorities), and efficient use of funds in the investment in new infrastructure. In this latter context it is noted that metro-rail systems while potentially having a role to play address only a small proportion of transport demands, at a very high cost premium, and in any case do not remove the requirement for a city to provide efficient on-street public transport. Cities such as New York, London, Paris, Rio, Sao Paulo and others which have high capacity metro systems also have high capacity bus services with priority on facilities. The need to tap the efficiency of the private sector has been demonstrated. The social dimension of urban transport was addressed both through a concern with the affordability of public transport and through a concern for non-motorized transport and walking.

That summary of the declared positions of the Government of India and the World Bank discloses a degree of agreement in principle, which should be the basis for a very fruitful collaboration in assisting the development of urban transport in India.

CURRENTLY BEING IMPLEMENTED TO ADDRESS THE URBAN TRANSPORT PROBLEM

The Government of India and many city authorities are dealing with the urban transport issue on many fronts. In Delhi, they are undertaking a major investment in the new Delhi metro system (with funding support from the JBIC). Several new flyovers have been constructed in recent years. Public transport vehicles have been converted to CNG. In Mumbai, the government is investing in a number of urban roads and suburban rail projects. The Mumbai government is also considering implementation of metro system. In Chennai, there has been recent development of a section of elevated MRT system which is a continuation of the suburban rail system, ring roads have been completed and urban expressway construction planned. In Bangalore, there has been extensive discussion of the severe urban traffic congestion problem and urgent measures need to be taken to address this. The city has been discussing metro options.

On the environmental front several measures have been taken to mitigate adverse effects of urban transport on air quality. India now has switched to unleaded fuel. In Mumbai at least the sulfur content of diesel has been reduced to levels at which Euro 3 standards for vehicle emissions can be set. Delhi set an example by undertaking a comprehensive and far-reaching program of measures, of which the most publicized is the mandatory conversion of city's public transport to compressed natural gas (CNG) in 2000-2002. This clearly reduced the visual impact of "black smoke" in the major public transport corridors. The Supreme Court has now directed a number of other highly polluted cities in India to prepare "action plans" for addressing urban air pollution, incorporating many of the measures adopted by Delhi.

But much remains to be done. There remains a need to address the content of sulfur in diesel more generally. With the dramatic increase in use of two-wheelers (especially for example in Chennai), there is a need to introduce measures to ensure the switch of motorcycles from two-stoke to four-stroke. One consequence of opening up of the automobile market to international competition coupled with the introduction of increasingly tighter emission standards has been a shift from two-stroke to four-stroke engines among two- and three-wheelers. According to the new vehicle sale figures, the sale of four-stroke engine two-wheelers increased from 21 per cent in 1997-1998 to 79 per cent in 2000-2004, with a corresponding decrease in the sale of two-stroke engine two-wheelers (SIAM 2004). There is a need to speed this up to only four-stoke being sold in India and to introduce measures to switch older in-use vehicles to four-stroke.

So far the World Bank contribution to urban transport in India has been relatively limited, despite the very large commitment to inter-urban road development. The Bank is currently providing funding support to the US$945 million equivalent Mumbai Urban Transport Project (Cr. 3662-IN/Ln. 4665-IN) which supports expansion of the suburban rail system, development of key connecting roads, and pedestrian facilities. In Chennai a further US$150 million equivalent of urban roads are proposed for development under funding of the proposed Tamil Nadu Urban Development Project III. Urban road construction in Bangalore will be included in the proposed Karnataka Municipal Development Project.

PUTTING POLICY INTO PRACTICE – THE CASE OF CHENNAI AND BANGALORE

Given the extreme need for immediate infrastructure investments there has been an ongoing investment in urban roads. In moving forward these type of investments will need to give greater support to the implementation of the urban transport policy. While Indian cities have developed comprehensive urban transport strategies which highlight the decline in public transport due to lack of attention to provision of facilities for public transport, there is little on street attention to the need for public transport priority in any of the urban road projects. Similarly, despite the discussion of non-motorized transport, there is little attention to the needs of the walking public through proper sidewalk provision (for example, in Mumbai 40 per cent of commuter trips are by walking).

With the intent of moving forward on the discussions on possible further World Bank support to Urban Transport the Bank undertook a review of the urban transport situation in two cities, namely Chennai and Bangalore. The reason for the focus on these two cities was that there is ongoing discussion of support to urban development projects namely the proposed Tamil Nadu Urban Development Project III and the proposed Karnataka Municipal Reform Project which could provide a forum for discussion of development of urban transport projects and policy reforms.

In the past 15 years, both cities have experienced a combination of population, economic and spatial growth that is placing a tremendous strain on their public infrastructure and services. Bangalore metropolitan area has a population of 5.7 million and is growing at 4.9 per cent per annum, while Chennai area has a population of 7.5 million, with a slower growth rate, just under 1 per cent. Motor vehicle ownership is increasing very rapidly; Chennai already has 324 vehicles per 1,000 people and Bangalore has 298 vehicles per 1,000 people. Motorized 2-wheelers are the main growth category, with more than one million registered in each city compared with about a quarter million cars. Despite this overall mobility rates are low, just above 1 trip per capita per day, and the passenger markets are highly heterogeneous, reflecting great inequality of income and wealth. Walking and cycling account for 44 per cent trips in Chennai and 17 per cent in Bangalore, and more than 40 per cent of daily trips take place on public transport services. Chennai's public transport system is dominated by street-based buses (38 per cent of trips), but it also has three commuter rail lines and one urban rail rapid transit line in the making. Bangalore relies on street buses only, though for some years it also has been trying to acquire some form of higher-capacity, rapid transit system. All operators are public sector owned. Due to modal shifts to 2-wheelers, the trend for the usage of public transport services has not risen in the past decade, despite population increases and (suspected) higher travel rates.

Beyond the sheer scale and diversity of the demands posed by the growth in population and incomes would have proved taxing for most world cities, there are some local factors, all interconnected, that explain this unsatisfactory state of affairs:

- On the institutional side, the transfer of powers and resources from states to local governments has been slow. The political constituencies of state and local institutions being different, the continuing dominance by the state produces transport policies and investments not well aligned with local interests. Large-scale investments (elevated highways, ring roads) tend to get more attention than street maintenance.
- The proliferation of state and local institutions and parastatals is

unusually high, resulting in diluted regulatory and funding authority, and accountability for urban transport matters. Neither city has developed capacity for public transport regulation.

- The urban transport sector does not generate any revenue surpluses directly available at the local level. National and state taxes on fuel and motor vehicles are substantial, but only a fraction (25 per cent nationally) is returned to the sector, and then in a circuitous way. Public transport has traditionally been a subsidized sector. The bus operator in Bangalore has in recent years turned an operating profit, but not yet in Chennai, where cost recovery is about 90 per cent. Commuter lines in Chennai are deep in the red, with 50 per cent recovery of direct operating costs. Funds for current and capital spending come from state budgets (under severe pressure in both states), and from the central government, via Central Road Fund, the Ministry of Railways, and city-bound programs like the Megacities Scheme. Together with other factors cited here, this way of funding biases spending in favor of large investment projects, some with dubious rationale, while leaving large urban and social segments poorly served.
- The use of competitive mechanisms is underdeveloped, as is the reliance on private sector funding and the know-how. In fact, it is limited to outsourcing of bus services in Bangalore, contract-based street maintenance, and a budding effort to charge for on-street parking in both cities.
- A laissez-faire approach has been taken with regard to the allocation of street space between competing uses. The losers of this are: (a) pedestrians; (b) bicyclists; and (c) public transport vehicles.

Both cities appear to have formulated the urban transport problem as that of street congestion and low safety, to be addressed in four dimensions. The first is to improve traffic flow by intensifying traffic police activities in traffic management and law enforcement, linked with some corridor and intersection improvements. The second is to take strong steps to improve the supply-side of public transport services, though largely staying within the public monopoly paradigm. Both of these efforts were necessary and the results achieved are impressive, especially on reducing traffic accidents in Chennai and improving bus operations in Bangalore. The third policy instrument is the addition of massive new road capacity in the form of multi-grade interchanges, elevated radial roads, and ring roads. The fourth, similarly capital hungry, is to move public transport development off-street onto the rail tracks.

Strategically, this approach is supply-oriented, and traffic growth-biased. It conflicts with the principles outlined in the government urban transport policy statement in a number of ways. In the short term it neglects the mobility of low-income and poor travellers, especially the non-motorized ones. It does not involve any use of traffic restraint tools and hence leaves street-based public transport services (the work horse of the transport system) to the mercy of unrestrained competition from individual motor vehicles. Moreover, it favors the most capital-intensive public transport modes (metros and other urban railways) which may not be warranted by either traffic density and passengers' ability to pay, or their budget capacity to pay subsidies in perpetuity. In the longer term the emphasis on increasing road capacity encourages car-based urban development patterns. The actual policies, as opposed to the statements in principle, thus appear to be both socially regressive, and financially unsustainable.

A POTENTIAL ROLE OF THE WORLD BANK

The Bank's Country Strategy recognizes that in India, urban transport is a high-risk sub-sector with fragmented responsibilities, weak fiscal and implementation capacity of local bodies, and complex safeguard issues. It is also potential high-return, especially if an emphasis on management improvements and traffic engineering helps defer high cost investments in mass transit and flyovers. Given the risks and competing demands on Bank resources, Bank lending support will be through pilots incorporated into operations to support broader municipal reforms. This engagement will be selective, focusing first on investments that have short paybacks, such as traffic engineering and management, bus ways, and slum accessibility. Drawing lessons from these pilot engagements, the Bank will thereafter seek opportunities for scaling up, IFC may also provide support in the area by investing in infrastructure development companies that are constructing and operating urban transport infrastructure.

Reflected in the Country Strategy are agreed "Guidelines for Bank Lending in Key Sectors", including in Urban Transport where instruments would include Analytical, and Advisory services (AAA) and investment lending limited to a few major urban centers and contingent on:

- Existence of a statewide urban policy aiming to clarify roles in urban development (including transport) and to enable ULBs to become financially viable;
- Quality of municipal management, including capital planning and budgeting, financial management, revenue administration, and financial performance;
- Willingness to prioritize investment using economic criteria; development of a sound urban transport development strategy and investment program, commitment to the introduction of modern traffic management and enforcement; and
- Commitment to institutional reforms required for citywide transport management.

The involvement of the World Bank has several beneficial prospects. First, its direct engagement in the growth-equity rebalancing will provide an added weight to the equity camp, much needed in these growth-dominated cities. Second, Bank loans can fund the planning effort for strategy development, and –through stringent engagement and selection criteria—ensure that some of the more difficult policy and investment shifts are tried, evaluated and refined. Third, the implementation of thus selected projects would re-direct immediate benefits to social sectors hitherto neglected in the current transport strategy. Fourth, given its urban and transport operations in the two states, a program approach is feasible.

The table below shows a hierarchy of 8 project types defining an exhaustive agenda of policy initiatives and investments. The current series of Bank-funded urban and transport projects in both Tamil Nadu and Karnataka, with their adaptive design and stress on local institutions and finance, provides a ready vehicle to test the three lower-rung projects/policy couplings. If these work well, free-standing urban transport projects in Chennai and Bangalore could aim at one of the higher-rung operations. A project to finance a rapid busway corridor (even a network) would be of highest priority in either city, because of its truly strategic investment and regulatory aspects. Proposals for bus-based rapid transit, in the form of sketch plans and outline cost estimates, are said to have been tabled in both Bangalore and Chennai, and could be built on readily and rapidly. The next three rungs (primary roads, commuter rail upgrading, and a metro line or metro access facilities) are project possibilities for the medium-to-long term, to be considered only if the strategic change has occurred.

URBAN TRANSPORT IN BANGALORE AND CHENNAI

OBJECTIVES AND CONTENT OF THE REPORT

The cities of Chennai (Tamil Nadu) and Bangalore (Karnataka) are mounting major efforts to deal with urban transport problems generated by exceptional rates of demographic, spatial and economic growth experienced therein over the last decade. The World Bank has a long history and a current presence as a partner in development endeavors of these cities and states. The areas of the current Bank activity include both urban and transport projects, but not specifically focused on urban transport. Given the perception of a growing importance of urban transport activities in both growth and poverty dimensions, an expansion of the Bank's assistance into this field, in the form of advisory and lending activities, is now being considered by all concerned parties.

The report in hand is intended to facilitate the discussions in this context, by providing an external angle of the urban transport problems, prospects and possible ways forward.

The main body of this report: (i) provides a brief diagnostic of the urban transport infrastructure and services in Chennai and Bangalore; (ii) identifies the underlying strategic issues; (iii) proposes an amended strategy, and (iv) outlines an agenda for the involvement of the World Bank in the short-to-medium term. The case studies of urban transport in Chennai and Bangalore, on which the main report is based, as well as a bibliography, are provided as attachments.

The report is a first-cut attempt to understand and address a complex subject. It is based on a brief field visit and desk research, both of which have disclosed serious lacunae in data. Other limitations have to do with a narrow focus on urban transport, adopted to make this initial attempt doable. For example, the report does not touch on the environmental aspects of urban transport, even though vehicle-produced air pollution is a major and increasing problem in both Chennai and Bangalore, indeed in all urban India. This omission is not likely to invalidate the proposals made herein, since they focus on potential increases in public transport patronage and on traffic restraint, both of which are unequivocally beneficial with regard to emissions. Conversely, the most important decision variables from environmental point of view (re vehicle emissions and fuel prices) apply at any level of modal split. A more serious limitation is that the report stays away from urban planning, land markets and municipal funding issues. Major analytical work is being done by the Bank in these areas, and its results are being incorporated into the design of lending operations. In the next stage of the work on urban transport, stronger links will need to be established between this subject and that of local government organization, funding and planning processes.

THE BACKGROUND

Transport Demand Characteristics

The main features of Bangalore and Chennai are shown in the following boxes. The two cities have similar population "masses," just above 4 million within city boundaries and about 7.-7.5 million in the urbanized area. Chennai is much more dense, but Bangalore is growing at a much greater rate (4.9 per cent per annum in the 1990s). Chennai is a long-established port city, with two adjacent centers also of older vintage - the traditional commercial hub next to a pre-independence administrative and military complex. Development spread from these centers and the port along a few major road and rail radials.

Its industries include petrochemicals, machine manufacture, and automotive equipment (both cars and rail rolling stock). Bangalore is land-locked, but at an important cross-roads of state/national roads and rail lines. Better known in the past as a city of gardens and lakes, whose moderate climate attracted pensioners and vacationers in large numbers, it has become a world-known center of information (software) technology, a synonym for outsourcing services for the U.S. and Western European countries. Bangalore's economy is much broader than its international image: most employment is in fact provided by trade and commerce (60 per cent in 1995), and manufacturing (37 per cent). Traditional activities like silk weaving and garments are also vibrant. Though Bangalore also has inherited two strong centers, it is much more poly-nuclear than Chennai and its road system is more diffuse and complicated. This is in tandem with the fact that rail lines entering Bangalore were neither designed nor operated to cater for urban and regional traffic, so the city's growth and mobility patterns have been very much road-dependent. Chennai's transport system, though greatly road dependent, also leans heavily on its commuter rail services and (soon) on its first urban rail line, now only open on a short link.

Both cities have in the last 20 years experienced a combination of demographic, spatial and economic growth that has catapulted them into the forefront of India's great jump forward. These same processes have placed a tremendous strain on their public infrastructure and services. For transport management and planning purposes, it is of essence to understand the divergent patterns in population, location and income changes. Economic growth has raised incomes of a large number of people and their expectations as to the services they deem essential. In the transport dimension, the most visible impact of rising incomes is accelerated motorization (vehicle ownership and use), accompanied by a shift from public transport services to individually or company owned vehicles. In the spatial dimension, this means an increase in the degrees of freedom to locate residences. At higher income brackets, this typically means a choice of more distant spots of greater environmental and other types of amenity.

Motor vehicle ownership in Bangalore and Chennai has been increasing at unprecedented rates, between 10 and 20 per cent per annum. The current ownership level is about 324 individual passenger vehicles per 1,000 population in Chennai, and 298 in Bangalore. These are high rates, similar to those in the wealthiest cities of Eastern Europe and common in Western Europe, but at vastly lower level of incomes than in Europe. The explanation for this seeming anomaly lies in the structure of the passenger vehicle fleet. Motorized 2-wheelers are the main growth category, with about 1.1 million registered in Chennai and 1.2 million in Bangalore.

Cars are a distant second: about 250,000 are registered in Chennai and 267,000 in Bangalore. This motorization pattern is similar to that experienced elsewhere in South and East Asia, *e.g.* Hanoi and Ho Chi Minh City in Vietnam; Kuala Lumpur in Malaysia. The consequences of 2-wheeler primacy, while a boon for the mobility of many people, are unfortunately quite negative for traffic flow, safety and air pollution. In terms of relations between motorization and incomes, car-based motorization is linked to higher and high-middle income households (in addition to business owners). Motorcycles, on the other hand, are bought by low-middle and low-income households. From transport planning point of view, they are bought by households who are "normally" major users of public transport services. Just how deep down the income ladder is motorcycle ownership was illustrated 10 years ago in a survey of bus passengers in Bangalore: 27 per cent of households with monthly income of ₹.500 or less owned a motorcycle (71 per cent owned a bike). For monthly incomes in the range ₹.500-1,500, 47 per cent owned a motorcycle. These numbers must have changed significantly since 1993, but the main point has not: many bus users are not captive and make their modal choice on the basis of some calculus of price, travel time, comfort, convenience, etc.

This said, the split of daily travel by mode is still not dominated by motorcycles and cars, but by public transport services, walking and biking. According to admittedly aged surveys in Chennai (probably 1992, with modifications based on more recent small-scale surveys), walking and biking accounted for 44 per cent of all trips, and public transport modes carried 42 per cent. The share of cars (2.5 per cent) is downright minor in comparison. In Bangalore, where data are even weaker but of more recent vintage, walking and bikes carry about 17 per cent of all trips, and public transport carried about 41 per cent (up to 60 per cent of all trips longer than 1 km), and individual motor vehicles carry 38 per cent. Even after newer and better data adjust these numbers downward, the visual evidence of unrestrained dominance of 2-wheelers, 3-wheelers and cars on the traffic scene in these two cities is misleading. The bias comes from focusing the visits and surveys on major street traffic. Urban transport also takes place elsewhere.

One of the reasons for the importance of non-motorized and public transport modes in Chennai, and somewhat less in Bangalore, is that economic growth has left many people behind. The new wealth is in sharp contrast to concurrent poverty, with inherited inequalities deepened by the growth processes, or new ones generated by them as the migrants from the countryside pour into cities. The population growth has taken place largely at the low-income end of the economic spectrum. In spatial terms, many of the lowest income people live in informal settlements in peri-urban areas, in older city slums, or encroach any place where development by leapfrogging has left some land unused. It is not that lower-income groups have not benefited from economic growth. Many did, but growth for this stratum of urban residents is in the informal sector, low-paid and unstable jobs held by unskilled workers in construction, diverse services, and informal manufacture.

Different income strata have different expectations of the urban transport system. Those owning individual motor vehicles, be they households or businesses (the latter including freight vehicles) expect a good road system: well-maintained pavements, efficient traffic control, high travel speeds, easily available parking. Rising incomes have also increased service expectations of some public transport passengers, especially if they own or aspire to own a motor vehicle. They expect higher-quality services: easy access, a seat, high travel speed, air conditioning (especially in Chennai with its humid and hot climate). Since the majority of public transport services in both cities operate on city streets, public transport passengers are also interested in the performance of the road system, as are public transport operators. Finally, and certainly not the least important aspect, a good-quality road system and good-quality public transport services are essential parts of a "package" that Chennai and Bangalore offer to potential investors from outside, in competition with other cities in India and elsewhere.

Transport expectations of people at the low end of the income distribution are very different from those holding formal and/or better paid jobs: they rely on walking, some in addition have bicycles, and those holding or seeking distant jobs rely also on public transport services. This implies, first, the demand for a basic network of all-weather roads in the secondary and tertiary category, linked to the arterial road system. Second, it implies minimally-priced and easily accessible public transport services. This simple 3-way segmentation of the travel market in Bangalore and Chennai does not capture the richness of what takes place on the ground. For example, the high-tech and engineering businesses of Bangalore have quite different transport habits and requirements than those than the traditional businesses, *e.g.* small-scale manufacture, silk weaving, commerce and services. The former are highly motorized, their job and familial networks are spread widely (well beyond Bangalore, in fact). As a caricature, it is this group that is conscious of traffic speeds and delays, and seeks flyovers, urban expressways and multi-level garages. The traditional businesses are more location-bound, with kin businesses locating in close proximity, and walking retaining importance for interaction between partners and with clients. These businesses may also be concerned for the ease and cost of longer-distance urban transport by motor vehicles, but within their large activity areas they do not seek to "reduce congestion" but thrive on it.

The Performance of Urban Transport Systems

How well are the transport systems of Chennai and Bangalore serving their diverse client populations? Answers should be sought both from the service providers (the supply side) as well as those for whom the services are provided (the demand side). A comprehensive and rigorous evaluation from the supply side is not available. The urban transport institutions in Chennai and Bangalore have not yet focused on the question of service to citizens in a systematic manner. The following evaluation is culled from various technical studies consulted for this report, complemented by visual evidence from a recent, but all-too-brief exposure to on-street conditions in the two cities. The overall conclusion is that the performance of urban transport systems in

Bangalore and Chennai leaves much to be desired across all economic and spatial strata. The worst off are the pedestrians in all parts of the urban areas, due to non-existent, broken-down, and/or obstructed sidewalks; large height differences between sidewalks and frequent driveways/alleyways; danger at street crossings and distance between crosswalk locations; and flooding in monsoon seasons.

The next on the list of poorly-served travellers are by bicycle riders, who have few exclusive-use lanes while gradually being pushed out of busy roads by motor vehicles, be these 2- or 3-wheelers, buses or cars. Traffic accident data from Chennai show that pedestrians and bike riders are second- and third-highest group among those killed in traffic accidents, with 190 and 126 killed in 2001, respectively (topped only by 208 dead riders/passengers of 2-wheelers). Traffic studies cite poor condition of pavements (30 per cent of Bangalore's road network is in that shape), low travel speeds (down to 10-12 km/h), high intersection delays, and poor or non-existent parking facilities. Traffic accidents are high at about 50 and 40 per 10,000 registered vehicles in Bangalore and Chennai, respectively, with about 700-800 fatalities (Bangalore is responsible for the upper range).

Bus services are infrequent and slow moving; buses are hard to get on/off, overcrowded (up to 150 per cent of the nominal capacity), with uncomfortable ride, and polluting. Suburban rail services have low frequencies, and difficult access to/from stations. These generalizations apart, a 1997 survey of MTC passengers in Chennai found 75 per cent satisfied with service frequency, 80 per cent satisfied with punctuality, 89 per cent satisfied with reliability, 93 per cent satisfied with safety, and 89 per cent satisfied with vehicle condition. The lowest score (48 per cent) was on "route condition" which probably refers to the road condition and possibly traffic delays. The same survey covered some potential and/or ex-passengers. The ranking of "push-away" factors was as follows: low travel speed, lack of punctuality, poor connectivity and low frequency.

Are public transport services affordable? A simple analysis of travel fares and passenger incomes for CMTC (Attachment I), based on the price of the monthly fare, concluded that bus fares were onerous at monthly household incomes of less than ₹.1,000 (roughly 10-13 per cent of passengers). At an income of exactly ₹.1,000, a monthly bus pass accounts for 14 per cent of the household income for a 10-km trip by one person, and 26 per cent for a 30-km commute. Commuter rail monthly passes were significantly more affordable. At ₹.2,500 a month per household, a monthly bus pass for one person would be under 10 per cent for most distances, and rail passes were half that. The conclusion is that fares are set at levels acceptable for a majority of passengers. Answers from the demand side come from two recent surveys. These covered several cities, including both Bangalore and Chennai. A 2003 Confederation of Indian Industry survey of urban populations in Southern

India showed 90 per cent dissatisfied with roads, and 58 per cent dissatisfied with public transport services. It is noteworthy that 65 per cent of the respondents were willing to pay higher public transport fares to get more comfort and frequency, and 89% of the respondents were willing to pay for good-quality toll roads. A 2003 study by the National Association of Software and Service Companies, done to evaluate the relative attractiveness of major Indian cities from IT business point of view, cited Bangalore's "weak public transport infrastructure (that) resulted in many people buying their own vehicle" and generally low infrastructure availability. The same study also cited Chennai as lacking in infrastructure. In other words, the dissatisfaction with infrastructure in Bangalore and Chennai is shared between the population and the business community. The evidence of merely two surveys cannot be taken as conclusive. Still, this is a serious situation since both cities perceive their chances of continued economic growth hinges on having much better infrastructure and services then at present, not to mention the satisfaction of their own citizens.

URBAN TRANSPORT ISSUES

The unfavorable evaluation of urban transport performance in the preceding section may be seen as unfair by those actively involved in the operations and planning of transport systems in Bangalore and Chennai. After all, major efforts have been made in both cities. In Bangalore, the last 6 years have seen an impressive revival of BMTC, including fleet renewal, increased punctuality, and lower number of breakdowns. All productivity indicators are up and the company has been making a profit for several years in a row. As for traffic congestion, there have been major road improvements, including an Outer Ring Road, the gigantic 5-km Hebbal flyover (the largest in India), and other smaller flyovers and underpasses at worst-congested intersections. More multi-grade projects are under construction and/or being tendered. The work on building the new international airport has started, and its road connections will be much better than is the case with the current airport. Chennai has constructed an Inner Ring Road and started on the Outer Ring. The most important radial roads in the city have been widened, and some have included pedestrian underpasses, and separate lanes for pedestrians and bicycles. Many intersections have been improved. A major effort was put in traffic law enforcement, lowering traffic accidents from a high of 5,280 in 2001 to 3,680 in 2002, and traffic deaths from 680 to 485 in the same period. New bus and truck terminals have been constructed. Phase I of the rail-based Mass Rapid Transit System (MRTS) was placed in operation in the late 1990s, and a gradual progress of gauge conversion has already made possible serving cross-radial trip ends without transfers. The completion of Phase II of the MRTS is imminent, creating a rapid urban railway of about 20 km which complements the existing suburban rail system.

While acknowledging that valiant efforts have been made in both cities, and real improvements have been achieved, it is clear that the efforts have not sufficed to keep up with loads and expectations generated by the demographic and economic growth. Neither financial nor institutional capacity of state and local governments were up to the task. In addition, some questionable policy and investment choices have been made, and others were left untouched. The rest of this section brings out the major among these factors, choices and underlying issues. Since all these are strongly interconnected, the order of presentation is to start from the most general factors. The working hypothesis is that the ensemble of state/city institutions in charge of the urban transport systems (with links to national institutions) are supply-focused rather than demand-focused. The resulting policy orientations and decisions on how to spend available funds have left large economic and spatial segments poorly served, and have not been as effective as they could have to make these cities competitive.

Finances

The structural problem with urban transport funding, which Bangalore and Chennai share with other Indian cities, in fact with many cities the world over, is that the sector does not generate any surplus revenue directly available to those who regulate, operate the transport systems and plan their development. Thus a growth sector (*e.g.* demand for roads) in an economically strong local environment (cities that are their states' and the country's leaders) cannot get an adequate supply response. In public transport services, the Bangalore Metropolitan Transport Company (BMTC) has started to generate revenue surpluses, but this has yet to be enough to upgrade the company's fleet for a visible rise in the quality of service. In Chennai, both the MTC and the commuter rail lines generate losses, cited above as ₹.1,347 million (US$ 28 million) in last year, and the prospect is that these losses will increase considerably when the MRTS Phase II becomes operational. The chances that fares could be raised in a significant way are not high.

On the road side, vehicle owners generate large revenues, through a state and national system of vehicle and fuel taxation. The states' taxes focus more on vehicle ownership, while the national tax is somewhat more use oriented. Most of the proceeds, however, are treated as general taxes: road sector expenditures are only 25 per cent of the total amount collected in road user taxes. The stress on vehicle taxation rather than fuel taxation is unfortunate, since it tends to reduce the potential of road use fees as an instrument for demand management. Moreover, the agency chain between what a vehicle owner in Chennai pays in vehicle and fuel taxes and what comes back to bear on road maintenance, traffic control, road rehabilitation and expansion in Chennai is quite long and indirect. In short, there is no close correspondence between increased demand for road space by motor vehicles and resources available to respond to that demand. Funds come to the urban transport sector in a variety of ways, from the state budget, from the Ministry of Railways budget (Chennai only) and through various national programs like the Megacities Scheme and the Urban Challenge Fund. While this is not an uncommon approach to urban transport funding, it is not well suited for a situation where an urban economy is stronger than its state's and its country's. Illuminating examples of a different approach, where locally generated funds are at immediate disposal of local institutions, accountable to local constituencies, include that of urban roads in Oslo and Bergen (Norway) and public transport systems in French cities outside Paris.

Institutions

The process of transferring the jurisdiction and resources from state to local governments, in line with constitutional reforms of 1992, has been slow, though accelerating in recent years. Municipal Corporations in Bangalore and Chennai are incomparably weaker in both authority and staff capacity. Their resource generating capacity is quite limited, the majority of funds coming in as transfers from their states. The capacity of smaller local bodies, outside the city limits but within the metropolitan area, is correspondingly lower. Given the joint nature of much of the transport infrastructure and services, the State Governments are de facto metropolitan governments. This would not be necessarily problematic if the distribution of political power (and therefore accountability) in state legislatures reflected the weight of large cities, their population and economic output. This has not been the case in either Karnataka or Tamil Nadu, at least not as far as the number of deputies in state assemblies is concerned.

There are several essential aspects in which the distribution of power and accountability between state and local government institutions affect urban transport matters. Risking a broad generalization, state transport agencies have an "aggregate" approach to the sector and ally themselves with big actors in the road and/or rial construction industry and others. This tends to lead to a preference for larger-scale investment projects, such as fly-overs and elevated roads in Bangalore, or even the MRTS in Chennai. City governments, council members as well as the bureaucrats, tend to be more responsive to local economic interests and local voters (including low-income populations). Whether this would also make them follow equitable and efficient urban transport policies has yet to be tested.

Reflecting the state/local split, neither city has vested the prime responsibility for all aspects of urban/metropolitan transport in one institution. Pieces of decision authority, control over resources and accountability are spread widely between state governments, local governments, and state and national parastatals. It is readily acknowledged that some fragmentation is both necessary and unavoidable. But, at any given level of fragmentation, there should be stable umbrella arrangements to coordinate various institutions. This is not the case here. In Bangalore, the fragmentation is truly extreme: in addition to state and city governments, plus local bodies outside Bangalore Corporation limits, plus two metropolitan area development authorities, the State has set up special-purpose parastatals (Bangalore Mass Rapid Transit Ltd., Karnataka Road Development Corporation, Karnataka Urban Infrastructure Development and Finance Corporation, this last a nodal agency for the Megacities Scheme) all of whom pursue some urban transport activities. The State has attempted to overcome the fragmentation by creating ad hoc bodies, such as Agenda for Bangalore, Transport Advisory Forum, and Task Force for Traffic and Transport, but these appear also to hold merely pieces of the pie. Bangalore Development Authority has no transport group, apparently no transport professionals at all. Indeed, its charter does not include transport planning. The last study with a comprehensive coverage was done long ago. In this forest of institutions, no single body appears to make comprehensive policy or medium-to-long term investment plans.

In Chennai, the situation is somewhat better. The charter of the Chennai Metropolitan Development Authority (CMDA) includes transport planning and the institution has a history of involvement with this subject, a team of experts and a well-developed network of local consultants. What Chennai lacks, and CMDA is not authorized to do, is public transport regulation. This subject may not have mattered in the past, but it does now.

The complicating aspect in Chennai is that the commuter rail services provided by Southern Railway network of Indian Railways play such a vital role in metropolitan transport. Service levels, prices and expansion plans of the commuter rail lines and the new urban railway (MRTS) are decided by different people than those for the bus system. This situation has multiple aspects. For the State of Tamil Nadu and the local governments in the Chennai metropolitan area it is advantageous that Indian Railways provide commuter rail services without any financial input from the state/local level. The gap between fare revenues and direct operating costs of these lines is about 50 per cent, amounting in 2001-02 to ₹.834 million (US$17.4 million). This compares to ₹.512.7 million (US$ 10.7 million) received in the same year by the CMTC, as a compensation for non-economic fares and services.

On the negative side, the state and local governments have little leverage in situations where the interest of Indian Railways' main lines of business diverges from that of the area's population. This works in the opposite direction as well, in that local government have had little incentive to organize things so as to maximize the ridership on commuter rail lines. In fact, some important decisions may have gone awry because the costs and benefits fell on different parties. The MRTS is a case in point. Phase I of the system was built with the federal funds (the State of Tamil Nadu contributed some land) and its large operating deficit has been met from the Railways budget. It is evident that Phase I has been nothing short of a functional and financial failure (carrying 9,000 passengers per day), made even worse by the CMTC running competitive bus lines. Had the funds used for the MRTS been available to spend locally, with operating subsidy also being a local responsibility, would the MRTS have been built? This said, MRTS Phase II is being built with 2/3 state participation, already a discipline-imposing move. The next step in this process is likely to be a transfer of the operating subsidy load onto the state government.

REGULATORY POLICIES IN URBAN PUBLIC TRANSPORT

Historically, state transport undertakings (bus companies) have been the prime providers of public transport services in most Indian cities, including Bangalore and Chennai. Fares have traditionally been set low by state authorities to permit travel by low-income citizens, especially those covering long distances. The inability of the state to pay fair and regular compensation, interacting with inefficiencies on the supply side stemming from the nature of public monopolies, chained public transport services to a low-service, low-priced equilibrium. A traditional and entrenched focus on production rather than service, rigidities regarding staff levels and remuneration, and low financial capacity combined to create a formidable barrier to change. With its ups and downs, this approach was acceptable while a great majority of passengers were captives, interested mainly in low fares. Greater incomes in the 1990s and an increased affordability of motorized 2-wheelers resulted in a large loss of public transport passengers, a process which is still underway and may acquire crisis proportions.

In Bangalore, there was a rise in private buses, as businesses moved to ensure that their employees came in on time and in comfort. Raising the level of public transport services therefore became essential. Since the public sector alone was not seen up to the task, the 1988 liberal legislation opened the door to private transport operators. What the legislation failed to do was to create a regulatory apparatus on each of the three levels of government, capable of dealing with a mixed public/private market so that the ensemble would evolve in the public interest. Very high levels of traffic congestion, pollution and safety hazards experienced in cities like Kolkata have demonstrated the dangers of un-restructured public sector combined with un-regulated private providers of public transport services.

The response to these changes in Bangalore, where the level of services by Karnataka State Road Undertaking had hit the bottom, was not to deregulate but to "cure" the public monopoly. This was done through a combination of actions, some on the relation state-company, others company internal. In 1997, Bangalore MTC was separated out of the state-wide company, and its organization re-structured, removing one layer of management. A fare adjustment formula, based on major input costs, was introduced, putting an end to the previous practice of fare approvals arbitrary in both scale and timing. An internal improvement program, focusing on both staff and management conduct, was implemented. The use of information technology was increased. The last but not the least is that BMTC opened the door to the private sector through outsourcing, even in its main business line – transport services. This consists of a "kilometer scheme" whereby private operators compete on gross cost basis to serve specific routes. In 2001-2002, close to 300 private buses were in operation, equivalent to about13 per cent of the BMTC's fleet. The sum of these efforts is evident in all technical performance indicators (fleet availability and utilization, passengers carried per vehicle, number of breakdowns, etc). It is also evident in its financial performance: the loss of ₹.78.2 million (about US$ 2 million) in 1997-98 turned to a small surplus of ₹.39.6 million in 1998-99, rising to ₹.267 million (US$ 5.6 million) in 2001-2002. The situation in Chennai had not been as dramatically bad as in Bangalore, so changes have also been less striking. CMTC has increased its cost recovery to 80 per cent in 2001-2002 and 90 per cent in 2002-2003, and the compensation payments have visibly increased over the last 5-year period, greatly improving the company's financial position. CMTC also is trying to introduce outsourcing of transport services to transport operators, but this has been challenged by the unions and the matter is in courts.

Missing from the above account are two essential variables. First, are MTCs in Bangalore and Chennai cost efficient? This question, politically very sensitive, has yet to be tackled. By international standards, both companies are overstaffed (more than 6 staff per vehicle in service). The average staff cost per month (about ₹.10,000) is in excess of what the majority of MTC's passengers receive. Second, what has been the impact of changes in the companies' performance on the service quality offered to passengers? Annual reports of both companies reflect very little interest in this subject. The performance indicators, other than the total number of passengers, are all supply-related. This may have been a normal and acceptable approach when most passengers were captives, but not when more than a half of them already own 2-wheelers, not to mention those who have already given up on bus services.

The essential remaining question is this: can the current regulatory arrangement, a public-sector monopoly, with an outsourcing complement, produce the cost efficiency and service levels to make this mode competitive with individually owned motor vehicles? A clear and promising option is to move toward a market-based arrangement, by separating regulatory and service planning functions from the provision of operations, organizing the latter through the medium of competitively awarded service contracts.

A similar dilemma has to do with the organizational status of commuter rail lines in Chennai, with the added complication that the current public sector owner is not the State of Tamil Nadu but the nation (through Ministry of

Railways). While the nature of competition available to with rail-based lines is much more limited than with street-based buses, the potential of service concessions is real.

The Fare/Quality Nexus

The co-existence of large "captive" and "choice" markets for passenger transport services, and the growth of the latter in proportion to the economic growth in cities places urban transport regulators in a dilemma. Keeping the fares low to assist low-income and poor travellers creates pressure on the budgets available for subsidies and involves a leakage of benefits to better off passengers. The lower the fare, for a given level of service, the higher the subsidy load becomes and so does the leakage. Conversely, for a given fare, increasing the level of services will also increase the subsidy load. Rail-based modes are especially sensitive to this, due to rigidities of large fixed costs. This in part explains exceptionally low fares on Chennai commuter rail lines.

The practice in both Chennai and Bangalore, low fare and a low level of service, has produced a flight to 2-wheelers. This in turn has produced a very heavy load on the road system. Both companies have introduced differentiated services (*e.g.* express and skip stop), and commuter rail in Chennai has different classes, to try to capture the quality-seeking passengers. Still, the flight continues and it will intensify if the current pace of motorization continues. The fare/quality issue has yet to be tackled as a strategic matter in either city. Proposals to increase fares have been made, but arguments for doing so were limited to the finances of public transport operators. A full argument would include the predicament of lower-income and poor travellers. This would allow a full range of options to be considered, not just in the fare and service quality dimension, but also regarding the regulatory framework and the approach to social assistance. In other words, certain "informal" public transport modes may be better suited to serve low-density, low-income communities than the conventional ones. Also, direct financial assistance to poor travellers may be "cheaper" than keeping fares low. At this point, it would be difficult to have such a consideration, since demand-related data are so inadequate and the relevant technical skills are in short supply in the state and local institutions.

The Allocation of road Space

The subject of road space is a frequently visited one in the Indian urban transport context. It is most often argued that the available street space is much too low in all large cities except Delhi. This position is then used to argue not only for widening and building more (elevated) roads, but also for the construction of off-road public transport systems, be these metros, sky buses, etc. Other authors argue that the road space is not a problem, but its management is. In all likelihood, both parties are right. The street space needs to be managed much better, and building new roads and exclusive-track public transport system is warranted in cities which are coping with traffic loads for which their networks certainly were not designed. The essential questions are, of course, who is going to get the street space available at present, how much new road space is to be provided and which off-street systems are going to be built.

The way this subject is approached in both Bangalore and Chennai has been to: widen the existing roads to a maximum possible, leaving a meager sidewalk width for pedestrians; and apply a laissez-faire attitude to what happens in traffic lanes. What happens is of course that (a) motor vehicles push off the bicycles, and (b) public transport vehicles lose the battle with more nimble 2-wheelers and cars. In addition, parked vehicles generally are allowed to obstruct the moving lanes. Save for some prohibitions against the use of goods vehicles in certain hours, there is no policy of traffic restraint. This omission is deleterious from both fairness and efficiency point of view.

A special case of traffic restraint has to do with public transport services. No matter how excellent the supply side of public transport operations may be, the service will only have as much quality as the traffic conditions will allow. In both Chennai and Bangalore, this is truly a strategic issue. Neither city has introduced public transport priority measures on city streets, not to mention the creation of at-grade, exclusive-use corridors and networks for bus services. This is not for the want of trying by planners. In Chennai, a busway on Anna Salai was designed and made ready for inclusion under one of Bank-funded urban development projects, but was withdrawn. Only a short exclusive bus lane remains from this scheme. The 10-year investment plan for Chennai contains an elevated highway along Anna Salai, but not an elevated busway. In Bangalore, BMTC commissioned a feasibility study for a bus-based mass rapid transit system. The study, completed in 1999, identified a promising network of 20 bus routes, composed of a Syamese-twin central rings intersected by 8 radial routes. A pilot 12-km line from Jayanagar in the south to Shivajinagar in the north, was estimated to cost ₹ 394.9 million (US$ 8.6 million). This includes the corridor and depot infrastructure and 35 special-purpose buses. So far, there is no move from the authorities.

The consequences of this approach are negative for both street-based bus operations and for chances to acquire an off-street public transport system. When low-cost options for the latter are neglected or rejected, only the expensive ones stay on the table. At the very least, this means that fewer corridors can be provided with off-street public transport modes. The best available advice, based on comparative studies of strategic responses to motorization in many Asian countries, is that the provision of separate space for public transport vehicles and private vehicle restraint are crucial at an early stage of motorization.

Metros

The neglect of bus-based rapid transit modes in Chennai and Bangalore, indeed in India generally, is proportional to the affection for rail-based modes, especially metros. Rare is an account of urban transport in India which does not mention the Kolkata Metro and the Chennai MRTS, or more recently the Delhi Metro. This may have to do with the larger-than-life role that railways played in Indian history and a common association of metros with great cities of the world. The resulting bias has an operational form in the view that railways belong to the exclusive tracks and buses belong on the street, or to connect villages.

The history and the present of transport planning in both Chennai and Bangalore is replete with plans to build a metro or some kin form of urban rail. Chennai actually went ahead and built the first short section of the MRTS and is about to complete the second (a combined length of about 20 km). In addition, the city plans to continue the MRTS (in the circumferential direction), and place a rapid railway line in the middle of the Outer Ring Road. A metro in Bangalore was recommended as early as 1982, then again in 1983 when Southern Railway produced a comprehensive commuter rail development plan. Another study in 1988 (funded by the World Bank) focused on the commuter rail, whereas the next one in 1993 returned the focus to a 2-line metro. In 1994, the attention shifted to a light-rail-based, 6-route, partially elevated network.

This was to be developed as a private-public partnership, and operated on a concession basis. This project proceeded beyond a mere proposal, but stopped when the private partner (after more detailed demand studies) asked for a much higher public participation than initially proposed. Finally, in 2003, a new feasibility study proposed a 2-line metro (18 and 15 km), a cross-shaped system designed to connect all major rail and bus terminals, and most activity centers. It is estimated to cost ₹.49.89 billion (roughly $1 billion) in 2003 terms. The financial engineering would follow a successful approach used to build the Delhi metro, *i.e.* 33 per cent the state of Karnataka, 22 per cent national government, the rest to be borrowed long term from both domestic and external sources. In the fall of 2003, a feasibility study for another metro in Chennai, using the Delhi and Bangalore approach, was being considered by the Government of Tamil Nadu.

Without prejudice to any of the past or current metro proposals, two general issues are involved here. The first is that the attention to metros may be an obstacle to doing something tangible to improve the position of street-based bus lines, *i.e.* some combination of exclusive lanes with priority of passage at signals, and constructing bus-based rapid transit lines in one of many candidate corridors. The second is the approach to doing feasibility studies. Investments estimated to cost billions of rupees tend to be put forward with single-valued outcomes of major items, *i.e.* construction and operating costs, passenger volumes and revenues. The notion of risk is absent. This is troublesome, especially given the abysmal record on cost, construction period length and traffic forecasts in Kolkata and Chennai urban rail projects. Also, the studies do not focus on alternatives to the proposed system. This may have to do with a trend that all feasibility studies for large rapid transit investments (rail or bus) are done by promoters of various systems, rather than commissioned by the transport planning authorities from independent consultants, with safeguards written into the terms of reference.

The Underlying Strategy

Neither city has formalized a comprehensive urban transport strategy, linked to an urban development strategy. What underlies the ensemble of actions, plans and proposals cited in this report appears to be: negligent of pedestrians, non-motorized and local area travel; (engineering) supply-driven; overly accommodating to individual motor vehicles; conservative in public transport regulation; non-protective of street-based public transport modes; and overly focused on large-scale investments, rail-based public transport investments and primary roads, in apparent belief that these visible structures will increase the image of competitiveness of their city.

CITIES NEED A DEMAND-SEGMENTED, SERVICE-ORIENTED URBAN TRANSPORT STRATEGY

The two cities need a demand-segmented, service-oriented urban transport strategy, which would balance growth with equity concerns, with a strong but cost-conscious orientation in favor of public transport modes. The demand segmentation is meant to re-direct the attention to low-income groups and sub-areas, but it is equally warranted in public transport regulatory matters because of the increasing size of the "choice" market. Practically, this strategy would involve making the following progression of steps, from simple to the more difficult:

- Measure and evaluate the performance of the transport system, regularly, from the point of view of different groups. This would require a primary effort by the lead urban transport agency, to design the data requirements for different sub-sectors and agencies, commission an initial data collection effort, and maintain a data bank in perpetuity.
- Introduce road and street design standards and practices that are walk-and bicycle-friendly. This should start by including detailed instructions in the terms of reference for planning and design studies.
- Re-allocate the existing road space to provide substantial exclusivity and priority of use to public transport vehicles on arterial streets. The corollary of this is that general traffic would be restrained and parking would be controlled/priced. This would start by a pilot study

focusing on selected corridors and/or areas, to be followed by implementation and scaling up of the effort. Both design and implementing stages would involve the local government, traffic police, the transport operators, and the metropolitan planning agency. A substantial intensification of traffic and parking management activities would be required, which may lead to a re-allocation of traffic management functions between the traffic police and municipal administrations. The formation of strong traffic management units in the latter group will be necessary.

- Shift attention and resources to repairing and/or constructing anew secondary and tertiary urban road networks within low-income and poor areas, and connecting them to the arterial network. This requires a policy shift, to be reflected in the normal budgeting process. A link to items 2 and 3 above is needed.
- Address squarely the issue of public transport fares, subsidies and service levels, balancing social protection and modal split concerns, for all transport modes. This is a major lacuna in the present strategy. Corrective actions will require the setting up of a metropolitan transport regulatory authority, with a small professional support group, aided by external consultants.
- Implement a regulatory reform aimed at getting substantially higher-quality services and/or lower production costs (internal incentives for MTCs, a gradual move to competition; new organizational form for commuter rail and MRTS in Chennai). The cited regulatory group is a pre-requisite for considering options and implementing changes.
- Develop a market for public transport modes suitable to serve travel demands at the low end of the income distribution (this also may involve breaking the monopoly of MTCs). The cited regulatory authority is essential for this task.
- Introduce rigorous project evaluation for large projects, inclusive of mandatory options and risk-conscious analysis. This can start by carefully designed terms of reference and short-listing criteria requiring a much greater involvement of independent consultants.
- Focus on at-grade, bus-based rapid transit lines, with publicly-owned infrastructure and competitively awarded service concessions, (inclusive of feeder/distributor networks). A pilot project will be necessary to break through the long-held biases.
- Ensure that new primary roads include a provision for rapid public transport modes (no reference to a specific vehicle technology). This is already a part of some road projects (in Chennai), but so far has been biased in favor of rail-based systems.

How to move in this direction? The transition from a narrow, supply-oriented approach to a demand-oriented one is a formidable task. Three ingredients are essential. First is the political agreement with the strategy, difficult because the proposals run counter to pro-growth forces, unions, motor-vehicle owners and the formidable urban rail lobby. Second is a streamlined and strengthened institutional setting. For a start, this would involve the appointment of a lead urban transport institution in Bangalore and strengthening of the Chennai Metropolitan Development Authority. Next, it would involve the creation of a public transport regulatory authority, a policy making body whose technical support can be provided by a separate unit (as in Step #5 above), or by the lead transport planning institution. Also, as noted in Step #3 above, creating a strong traffic management focus group in the municipal engineering structure will be needed, with some realignment of functions of the Traffic Police.

The third ingredient is financial. In addition to current efforts to improve funding, budgeting and expenditure management of local governments, there is a systemic problem that transcends Chennai and Bangalore, indeed their states also. It has to do with the national approach to road user pricing and revenue allocation. The problem is to reduce the overlong agency chain between what is paid by local road users (a growth sector in two well-off cities) and the funds brought back to bear on the local transport system. There are several ways to do this. The most common way is to escape funding from general (national, state or city) budgets, by creating a closed loop from road user fees via dedicated funds to cities. A less common way, highly successful where it has been implemented, is to introduce local road charging systems, aiming for both revenue generation as well as demand management. Either way, the challenge is to create not merely urban road funds, but urban transport funds, open to all modes. Private sector funding has a potential as a complement, but the prime source of funds should be user-based and locally linked. This subject is currently beyond the decision making reach of cities, but it needs to enter the discussion agendas at all levels of government.

THE POTENTIAL ROLE OF THE WORLD BANK

The involvement of the World Bank may increase the chances for the development, formal adoption and implementation of the above strategy. First, its direct engagement in the politically difficult growth-equity rebalancing will provide an added weight to the equity camp, much needed in these growth-dominated cities. Second, Bank loans can fund the whole sequence from the design of new type of planning and investment studies, through project selection using stringent engagement and selection criteria, all the way to implementation and evaluation. The Bank's presence would ensure that some of the more difficult policy and investment shifts are tried, evaluated and refined. The implementation of thus selected projects would re-direct immediate benefits to social sectors hitherto neglected in the current transport strategy, which is one of the Bank's primary objectives. Fourth, given the Bank's long history of involvement and its continuing urban and transport projects in the two states, a program approach is feasible.

A hierarchy of 8 project types defining an exhaustive agenda of policy initiatives and investments, based on the preceding list of strategic moves. Lower-rung options represent small-scale departures from the current practice in the Bank-funded urban and transport projects in both Tamil Nadu and Karnataka. The follow-up projects, now under preparation, with their adaptive design and stress on local institutions and finance, provide ready vehicles to introduce and test policy "turns" in favor of pedestrians, NMTs, public transport modes, and low-income areas. If these policies take root, free-standing urban transport projects in Chennai and Bangalore could aim at one of the higher-rung operations. The highest-rung options are provided to illustrate what may be doable (and will become necessary) in the longer term.

A project to finance a rapid busway corridor (even a network) is deemed to be of highest strategic priority in either city, as a vehicle to tackle and resolve the underlying conceptual, funding, and regulatory issues. Proposals for bus-based rapid transit, in the form of feasibility or at least pre-feasibility-level sketch plans and outline cost estimates, have existed for some time in both Chennai and Bangalore. These require capital investments of under $10 million for pilot projects in single corridors. Such proposals could be developed and implemented readily and rapidly. The next three rungs (primary roads, commuter rail upgrading, and a metro line or metro access facilities) are project possibilities for the medium-to-long term, to be considered only if the strategic change has occurred.

6

Air Traffic Control Systems

As air traffic continues to increase, new technology will be needed to accommodate it. The interesting opportunities that new technologies will present for air traffic control (ATC) will be matched by the human factors challenges. Automated tools, already common in the cockpit, will become a necessity for air traffic control specialists and maintenance personnel. The nature of the jobs of controllers and maintainers will change with the tasks that they are required to perform and the tools that are available to them. The human factors challenge is to ensure that these tasks and tools are designed to be well-suited to the users. How well we meet this challenge will determine whether the implementation of a new system is a success, a struggle, or a failure. The purpose of this article is to explore lessons learned from the development and implementations of several systems in the United States. These lessons point to a process that can be used to help ensure that new systems are designed and implemented effectively.

The first step in the successful implementation of any ATC system is careful planning. This planning must include strategies for ensuring that systems are well-designed from a human factors perspective. Human error remains the most common contributing factor in aviation accidents and incidents. The initial investment in attention to human factors pays off first by capitalizing on the extensive body of information on factors affecting human performance that is available to the human factors specialists. Systems that consider the capabilities and limitations of the human operator in their design help to reduce the probability of human error (and limit the consequences of inevitable human error) and thereby reduce the program's technical and safety risks, lower implementation and life cycle costs, and increase the probability of program success. Early consideration of human factors issues will result in potential problems being detected earlier, and resolved earlier, than if human factors planning is delayed or non-existent. The earlier in the acquisition process that problems are identified, the easier and less costly they are to correct.

Early and Continuous Focus on Human Factors. The initial human factors investment for a program should be in the development of the mission needs statement. This defines why the new system is needed, what you need the system to do, and describes any known operational constraints. From this, the careful description of operational and human factors requirements can be developed. This process begins with the details of what you want the system to do and identifies the tasks that the operator (controller or maintainer) will perform and the information that the operator will need to perform these tasks. This careful description of the tasks and information required to perform them is the first step in identifying the human factors requirements that will need to be met by the system. This process should adopt a team approach that has both users and human factors specialists involved in the specification of the requirements, the evaluation of prototypes, and the operational testing. The success of any system is measured by how well it meets these requirements. The specification of user and system requirements involves careful consideration of:

- task requirements (*e.g.*, What does the controller or maintainer need to do with the system? What duties will the users be expected to perform concurrently with the new equipment? How will the tasks change with the new system?)
- operational environment (such as airspace characteristics; amount, type, and complexity of air traffic; local procedures, etc.)
- characteristics of the users (this includes understanding the skills of the users that must be preserved; and the practices, procedures, and equipment that the users are accustomed to)
- transition to the new system

The need for early and continuous consideration of human factors issues was a painful lesson first learned in the Advanced Automation System (AAS). This was an ambitious program to significantly upgrade the air traffic control systems used in the terminal and en route airspace. This included redesigning the controllers' workstations and displays, and the software to support them. In the document, "Lessons Learned: Human Factors in the AAS Procurement", Small (1994) states that, "Some of the difficulties with controller acceptance of AAS could have been alleviated by involving human factors expertise earlier and by integrating it more fully into the design process" (p.4).

Three years later, the same statement could be made about the Standard Terminal Automation Replacement System (STARS) system. STARS was designed to replace the current radar processing and display system in the terminal environment. This program was the first of its kind to attempt to acquire the system by purchasing commercially available, "off-the-shelf" (COTS) equipment (as opposed to paying for the development of a new system). The appeal of this approach was the expectation of significantly lower cost and less time required for implementation. Unfortunately, this approach was interpreted as being incompatible with a complete specification of human factors requirements. In additional to minimal human factors requirements, the initial human factors plan (dated 23 February 1995) acknowledged that, due to an aggressive schedule, there would not be time for any human factors design development, nor was a full scale human factors evaluation planned. This combination of minimal human factors requirements, minimal consideration of human factors issues and the deficiencies in human factors planning, proved to undermine the initial development of the STARS program. In his testimony to Congress, the Inspector General of the United States identified the Federal Aviation Administration's (FAA's) "decision to limit human factors evaluation" and the lack of a formal process to "identify, prioritize and resolve human factors issues as the system was being developed" as two of the shortcomings in the STARS acquisition program (testimony of Ken Mead, 30 October 1997).

STRUCTURED USER INVOLVEMENT

Another problem area in both the AAS and STARS programs involved the use of controller opinion in the design process. It is important for users to be involved early and continuously throughout the design and acquisition process. It is equally important that this involvement be structured and integrated with the involvement of human factors specialist so that systems are not designed solely by user preference. Users should have a well-defined role in each stage of the acquisition and their tasks should be clearly specified (such as the evaluation of a prototype through the use of a questionnaire). The developers of AAS found that the preferences of controllers often changed with the group of controllers (Small, 1994).

This should come as no surprise, since individual preferences are based on individual experiences (such as, the characteristics of the airspace that the controller is accustomed to). Furthermore, it is well known that performance and preference do not always match; we do not always perform better with the design that we prefer. Finally, the controllers who volunteer to be a part of such efforts are likely to have more experience, and be more skillful and technologically inclined than controllers who choose not to participate. It is difficult (if not impossible) to put these skilled professionals in the operational shoes of a less skilled controller. Yet, to minimize the probability of human error, systems must be designed for a below average controller on a bad day.

While the AAS program may be said to have suffered from too much information on user preferences, the STARS program suffered from too little, too late. As if in response to the AAS experience of the system developers having to chase the system requirements, the involvement of line controllers in the development of the STARS requirements and other stages in the acquisition process was minimal. This resulted in costly delays in human factors problems being identified and addressed. Structured input from a broad spectrum of users is a critical component in identifying potential operational and human factors problems.

Prototype Assessment

The value of prototype assessment should not be overlooked. While the value of the information obtained from such testing will be dependent upon the stage of development in which the prototype assessment is conducted, even a rudimentary prototype can point to features of the system and procedures (such as data-entry procedures) that are likely to induce human errors or be operationally unsuitable for other reasons. As the design becomes more sophisticated, prototype testing offers a preliminary look at whether the system is likely to be able to perform its intended function and meet human factors requirements.

The value of prototype assessment is that it gives these insights into changes to the system that may need to be made at a stage of system development where changes cost much less than they will later. An example of critical information that resulted from prototype testing can be seen in the early days of the Traffic Alert and Collision Avoidance (TCAS) program. TCAS is a cockpit display of traffic information that issues an instruction to the pilot when a maneuver is deemed necessary to avert an impending collision between aircraft. One of the early developmental versions of TCAS included negative resolution advisories (RAs) such as "Don't Climb" and "Don't Descend". Prototype testing revealed that pilots responded inappropriately (such as climbing in response to a "Don't Climb") 50 per cent of the time a negative alert was presented in the operational simulation. As a result of these tests, all negative RAs were eliminated. (Boucek et. al., 1985).

Thorough Operational Testing

Even extensive prototype assessment does not detract from the need to conduct thorough operational testing of the design that will be implemented. Careful and thorough human factors testing of a system can be combined with the formal operational testing, although the formal operational evaluation should never be the first human factors test that is conducted. Such testing is necessary to validate the preferences of user's and the best estimates of human factors specialists.

User consensus is never a valid substitute for objective performance data. Objective performance data must be collected in order to ensure that the system is ready and suitable for implementation. Any system test must be well-designed from a human factors standpoint. For example, if the evaluation includes a simulation, then the controllers chosen must be representative of the user population (and not chosen on the basis of seniority, for example) and the tasks included in the simulation must be representative of those that the user will need to perform with the new system. Guidance for ensuring that an evaluation is well-designed from a human factors perspective, along with guidance on human factors planning is offered in "Human Factors in the Design and Evaluation of Air Traffic Control Systems".

Finally, it would be a mistake to assume that the need for such testing is minimal because the system is operational in another part of the world. Air traffic control is not "one size fits all". Systems need to be suitable for the operational environment in which they will be used. This means that characteristics of the airspace (amount, type, and complexity of traffic, local procedures, etc.) and characteristics of the users (*e.g.*, their skills, knowledge, the type of equipment they are accustomed to) and the tasks (what do you expect the user to be doing in addition to using the new equipment) all need to be considered.

An Example

One successful program that followed all of these steps was the Center TRACON Automation System (CTAS). The operational need for CTAS was evident years ago as increases in traffic were able to be accommodated en route much more easily than in the terminal environment. Originally developed at NASA - Ames Research Center, the system shows the controller the best runway assignment, landing sequence, and other information. This system was developed by engineers, human factors specialists and controllers. There was extensive prototype assessment with operational controllers that led to many design changes. The result was a system that is currently operational at Dallas/Fort Worth airport and has already demonstrated an ability to increase capacity. Under controlled conditions during a test conducted in 1996, CTAS was able to increase aircraft operations from 102 operations to 120 operations per hour.

In addition to doing things right from a process standpoint, CTAS also serves as an example of well-designed automated tools for controllers. Far from being an automated system that requires controllers to "feed and care for" the system's data needs with little or no return on their investment, the CTAS tools proved user-friendly and provided controllers with guidance information that they can readily use. Anecdotal reports indicate that the controllers at Denver and Dallas/Fort Worth who have used it like it very much. Dick Swauger, the national technology coordinator for the National Air Traffic Controllers Association, says, "It's like having a top controller whispering in your ear...it makes good controllers better." (Perry, 1997, p.31) As CTAS is implemented at other facilities, the increases in capacity that these tools can support (along with the controllers' acceptance and support for this system) is likely to be realized as long as CTAS continues to provide useful and user-friendly tools.

SYSTEM INTEGRATION

With the independent development of systems and subsystems for ATC, system integration becomes a critical issue. Even within a single system, integration can become an issue. For example, developing a system and its back-up or different components of a system independently minimizes the probability of both sytems failing for the same reasons and helps to ensure that there is not a single point of failure. From an engineering standpoint, this approach is highly desirable. From a human factors standpoint, however, the approach can be problematic if steps are not taken to ensure that the interfaces of the two systems are designed to be compatible, if not identical.

With systems developed independently, the issues become even more complex. However, the same human factors approach that has been outlined here for the development and implementation of new systems can also serve as an outline to ensure the effective integration and compatibility of separately developed systems. An example of the use of this approach can be taken from the cockpit. In 1979, human factors specialists at Boeing, Douglas, and Lockheed (Boucek, et. al., 1980) looked at cockpit design within and across manufacturers. What they found was a an excessive number of alerts and warnings that pilots were required to respond to and a variety of important inconsistencies in cockpit design. They found differences in aircraft cockpits, both between those developed by different manufacturers, and among cockpits developed by the same manufacturer, that could result in "negative transfer", that is, pilot errors that would be induced in one cockpit by virtue of having extensive experience in another cockpit. They also found situations within individual cockpits that could induce pilot errors, such as an excessive number of alerts and warnings.

Having established an operational requirement for aircraft alerting functions to be more effectively integrated, Boucek et. al. (1980) then set out to determine the best way to design cockpit alerts and warnings to be consistent and consolidated. Human factors specialists worked with engineers to design and test prototype alerting and warning systems that would meet a range of key operational requirements, such as minimizing the number of aural alerts, providing the flight crew with an indication of the level of urgency, and fitting in the space available in the cockpit. There were many human factors issues that needed to be addressed in this endeavor. Some of them, were able to be answered from the wealth of human factors knowledge already available. Many others, such as specific formats for voice messages and whether a voice message should be preceded by a tone, had to be prototyped and tested. After a series of studies, the end result was recommendations for aircraft alerting sytems that would serve cockpit manufactures for decades (Berson, et. al., 1981). Rather than have hundreds of individual alerts and warnings, they would be consolidated into master warning or a master caution, depending on the nature of the information.

Another valuable conclusion of this work was that there should be two levels of information presented to pilots. One level of crew alerting is status information that informs the pilot of situations that are important (such as a possible engine fire), but require the pilot to stabilize the aircraft before responding to the alert. The other level is guidance information that requires the pilot to make an immediate control action. An example of this level of alert is the ground proximity warning system (GPWS) that requires the pilot to "pull up" immediately to avoid an impending impact with terrain. The nature of the task required by the pilot determines the characteristics of the alert. It is easy to see how these lessons learned in the cockpit - the need for consistency in, and integration of, alerts and warnings and the need for alerts to indicate their level of urgengy - are applicable to air traffic as the number and types of alerts and warnings for controllers increases.

These lessons learned, both in the cockpit and in ATC, present a powerful case for:

- Attending to human factors and integration issues early in the acquisition process,
- Obtaining structured user input at various stages of system development,
- Prototype testing, and
- Thorough operationally-oriented human factors testing prior to implementation.

Giving human factors issues their due consideration at all stages of acquisition can present complex organizational and managerial challenges. However, this investment is a necessary step toward making the most of the opportunities of the future without repeating the mistakes of the past.

7

Rail Transport Services

INTRODUCTION

Railway companies were set up and operated in the 19th century as competing private enterprises, even though at that time they already received practical government support (allocation of land alongside the railway lines being built in the United States, financing of infrastructure). In the early 20th century they gradually formed themselves into groups. In many cases, they were nationalized, partly because the arrival of road transport for freight and individual motor transport for passengers meant that railway networks showed a massive structural deficit. In colonial territories, the government authorities usually established and administered the railway company directly. In the area of urban rail transport systems, the majority of subway systems (which are included within the definition of rail transport in the Provisional Central Product Classification (CPC-W/120)) were also set up, financed and managed from the start by government authorities, particularly the local authorities, in view of the large investment required and the fact that such networks are rarely profitable.

A "classic" public monopoly therefore emerged (but this was not universal because in the United States, for example, freight transport companies were never nationalized), and this form of organization is still the most widespread. This model only started to evolve in the 1980s as a result of the pressure for deregulation, privatization and the granting of concessions, encouraged by the authorities in some developed countries on the one hand and on the other by the World Bank in developing countries and countries with economies in transition. The very low level of commitments on rail transport is undoubtedly due to a large extent to the difficulty of integrating the GATS concepts of multilateral liberalization into this traditional framework.

Within such a framework, rail transport is a natural monopoly with high infrastructure costs, indivisibility and substantial externalities. Because of these features, government authorities have imposed controls over entry, withdrawal, technology, operating practices, capital formation, pricing, frequency, the financial structure and accounting practices. Such companies are therefore vertically integrated; one single entity is responsible for the infrastructure, operation and marketing. The focus is on production and the company is centralized and tightly organized into a hierarchy, which has its counterpart in the high rate of trade union membership. A company may be State-owned (this was the case in Europe, Latin America, Africa and nearly always in Asia), or private (New Zealand, United States, Japan for certain companies). It may have national or regional geographical coverage (Japan, United States) with possible variations on sectorial monopolies (passengers in the United States, freight in Japan) or regional monopolies (passengers in Japan) that may sometimes compete (freight in the United States where several companies only involved in freight may compete on the same routes). It has traditionally been recognized that the advantage of this model of vertical integration is its capacity for planning, but practice has shown that its disadvantages are failure to respond to the market, sometimes questionable investment decisions, absence of any incentive to control costs and poor financial performance.

Since at least the 1970s, the growth in road carriage of goods and passengers and, to a lesser extent, air transport of passengers has significantly eroded the market share of the railways. To take an example, in the European Union the railway's share in terms of passenger/kilometres fell from 10.3 per cent in 1970 to 8.5 per cent in 1980 and 6.2 per cent in 1994, whereas the figure for automobile traffic rose over the same period from 75.1 per cent to 79.7 per cent and for air transport from 2.1 per cent to 5.8 per cent. The renaissance of passenger traffic due to high-speed trains has merely halted the trend without being able to reverse it. Future development of these high-speed trains remains dependent on finding the financing needed to build new track, and this is highly problematic, even when the private sector is brought in (cf. the financing problems of the Folkestone-London stretch of the Eurostar or the difficulty of finding enough private financing for the TGV projects in Texas, Florida and Australia).

This downward trend is even more noticeable in the freight sector, where the modal split share of railways in tonne/kilometres in the European Community fell from 31.7 per cent in 1970 to 24.9 per cent in 1980, 18.9 per cent in 1990 and 14.9 per cent in 1994, whereas the share of road transport rose from 48.9 per cent to 71.9 per cent. This trend is omnipresent, but the degree varies according to the country, the initial modal split ratio and the structure of the networks. In the United States, for example, rail transport is still well ahead in terms of volume (40.9 per cent in 1995) compared with road transport (28.9 per cent), but the growth trend is much lower than that for road transport (1970-1995: rail = +70.6 per cent, road = +123.4 per cent).

The trend is even more noticeable in terms of value because the nature of the goods transported has changed and there is a much larger proportion of light goods with a high unit value that have to be delivered rapidly. The railways' share has therefore gradually been confined to bulk and heavy traffic, although since the early days of containerization in the United States and later in Europe, they have tried to win back the high value-added traffic by establishing transnational freight corridors for container-only trains (for example, Gioia Tauro-Antwerp) with a guaranteed date of arrival and computerized tracking of the goods.

The development of combined transport remains extremely marginal, particularly piggy-back transport, and is highly dependent on subsidies or sometimes restrictive transit measures by government authorities, which utilize railways as an ecological and energy-saving alternative to the growing congestion on major highways.

In developing countries, railways are particularly important because they constitute the main form of mass passenger transport at a price accessible to the majority of the population. This explains why in 1995 China alone accounted for 18 per cent of passenger/kilometres carried in the world and India 18 per cent (for purposes of comparison, the figure for the 15-member European Community was 14 per cent, for Russia 9 per cent, and for the United States 1 per cent). Railway companies in these countries also face competition from road transport and problems in financing the maintenance and renewal of the infrastructure and rolling-stock.

A study by the United Nations Economic Commission for Europe in 1990 on "Europe's Railway of the 90s" (Trans/SC.2/172) gives a fairly good idea of the sector's economic balance at the beginning of the decade. On average, 50 per cent of the network and 50 per cent of stations deal with 90 per cent of the traffic, while the following passenger sectors show a negative performance: suburban services and slow trains (local or regional trains); for goods, the negative sectors are parcels, single wagons and piggy-back transport ("railroad highways"). In passenger transport, the sectors that make a positive contribution are high-speed trains and daily Intercity expresses, for goods, they are complete train loads with long-term contracts and complete trains of trailers and containers.

Many analytical accounting calculations have even shown that passenger traffic at best only exceeds marginal costs whereas freight traffic is capable of yielding net profits. This is confirmed in the United States where freight companies are private and usually profitable, whereas passenger traffic is the responsibility of one large State company, Amtrak, which receives a subsidy of 21 per cent (1991) and 12 private companies with concessions and State finance for up to 40-60 per cent. These figures also show that activities viewed as essential by the authorities and which have economic externalities such as suburban transport show a structural deficit (Japan does not fit this picture due to the density of its urban population and the property activities of railway companies). It can also be seen that half the network and stations only deal with 10 per cent of the traffic and their continued existence has more to do with regional administration and universal service considerations than economic ones. It is telling to note that projects carried out with World Bank support almost systematically include a component on closing lines that are particularly unprofitable.

All these characteristics can be clearly seen in the accounts of railway companies: for members of the United Nations Economic Commission for Europe, earnings from transport (passengers and goods) only represented between 20 to 70 per cent of the railway companies' income, with an average of 50 per cent. The rest was covered by compensation for the performance of a public service and for retirement indemnities, subsidies, and the State's coverage of the residual deficit. With regard to costs, personnel costs, including social security contributions, accounted for between 48 to 77 per cent, with a median value of 55 per cent. This ratio is declining with the gradual reduction in the number of employees. Lastly, the accumulated debt to finance investment led to financial costs of around 5 per cent of operating costs and this figure has since increased.

This unbalanced economic structure is even more marked in the case of urban rail transport, due inter alia to the high cost of the infrastructure and amortization. In 1989, for example, the average cost of building 1 km. of railway track on the surface was US$20-25 million while in tunnels a kilometre of track cost US$85-105 million. This explains why to date only 93 cities in the world have such a transport system (29 in the European Community, 14 in the United States, 9 in Japan, 4 in Central and Eastern European countries and 4 in China). The following table, taken from a World Bank study gives an idea of the structure of the accounts of urban rail transport companies in a representative sample of cities in developing and developed countries. It shows in particular (cf. penultimate column) that none of these companies cover their total costs by their operating revenue and, in general, the ratio is rather 20 to 40 per cent, irrespective of the size of the subway system or the technical choices involved. It should also be noted that old subway systems, where theoretically the cost of the building the lines has already been amortized, are not profitable either

Faced with a growing deficit, at the end of the 1970s, government authorities tried to improve the traditional rail model. This led in the first instance to free fixing of prices (for example, in the United States, railways were authorized to conclude confidential contracts with shippers; 1976 Railroad Revitalization and Regulatory Reform Act and 1980 Staggers Act), often followed by the creation of specialized departments (freight, passengers, long distance, regional passengers, maintenance) as profit-making centres each responsible for its commercial policy but sharing common costs with the other departments on the basis of analytical accounting (a typical example is the organization of British Railways from 1980 to 1994). In parallel with this development, there was growing concern to identify more clearly the public service constraints and a consensus emerged that government authorities, particularly local or regional authorities, should be called on to finance the obligation to provide regular train services in a clearly-defined way adapted to each situation: depending on the country, the financing could be in the form of concessions or contract-plans with a single operator.

In the course of a third phase, a new model separating operating activities from management and maintenance of the infrastructure gradually started to be imposed. The theoretical inspiration for this model is similar to that for the telecommunications, electricity and gas sectors. In the case of railways, the idea was that, even if the costs of the infrastructure could not be recovered, the gains in efficiency obtained by separating operating/traction activities, no longer hampered by the financial burden of the infrastructure, would alone justify the State writing off the debt incurred by investment in infrastructure in its profit and loss account. This model was imposed in the European Union by Directive 91/440/EEC and in this particular case was coupled with the intra-Community liberalization of traffic. This step towards the international liberalization of rail traffic was the first in the world and warrants a relatively detailed description.

Directive 91/440/EEC gives "international groupings" (of one or more railway undertakings in member States) access and transit rights in the member States of establishment of their constituent railway undertakings, as well as transit rights in other member States for the supply of international transport services among member States where the undertakings constituting the said groupings are established. Furthermore, individual railway undertakings, (excluding urban, suburban and regional transport) are given the right of access, on equitable terms, to the infrastructure in the other member States for the purpose of operating international combined transport goods services. Subsequent Directives defined the regime for the licences that need to be obtained in order to be considered a railway undertaking within the meaning of Directive 91/440 (Directive 95/18 of 19 June 1995) and the criteria for the non-discriminatory allocation of infrastructure capacity and the charging of fees (Directive 95/19 of 19 June 1995).

These regulations are too recent for it to be possible to assess their impact fully. Nevertheless, in the light of the trade initiatives that immediately followed, the impact appears to be significant. Some member States interpreted the Directive strictly and established paths restricted solely to groups of national companies baptised "freightways". These include the 17 Antwerp-Lyons paths, with extensions to Marseilles and Barcelona, on the one hand, and to Turin-Genoa-Milan-La Spezia-Giaoia Tauro, on the other, set up in 1997-1998 by Interdelta/Belitalia, a grouping of Belgian, French, Italian, Spanish and Luxembourg railway companies.

Other member States (Germany, Netherlands, Austria and Italy, together with Switzerland in this instance) have gone beyond the Directive and have put in place three freight corridors between Germany, the Netherlands and eventually Scandinavia, on the one hand, and Austria, on the other, this time called "freeways" because they are not only open to national operators but also to any recognized rail operators within the meaning of Directive 95/18. The two cross-channel paths set up by the SNCF, the French rail company, in collaboration with the private British operator, EWS, subsidiary of the American company, Wisconsin, can also be considered "freeways", but in part at least these paths are not covered by the scope of the Directives. Shipowners such as the American company Sealand, the British company P&O and the Netherlands company Nedlloyd have already joined with the Dutch railways, NS, to set up a company, ERS (European Rail Services), to operate trains on these freeways. NDX, which is comprised of the Deutsche Bahn, the NS and the mother company Sealand, as well as the American company CSX, is another example of these international groupings. Lastly, in spring 1998, Deutsche Bahn and NS cargo merged their freight activities in a transnational company called "Rail Cargo Europe".

In two successive White Papers on 30 July 1996 and 29 May 1997, the Commission of the European Communities proposed to extend access rights to all freight and international passenger services. It also proposed that "freeways", which would remain based on voluntary arrangements among infrastructural networks, be open to all licensed operators and be able to undertake cabotage traffic (which would be partially opened up in accordance with a staggered timetable). It also proposed that rail, road and river transport operators be given fair, equal and non-discriminatory access to freight terminals. These proposals have encountered strong opposition from certain member States and trade unions.

In several member States, access has been much more open than that envisaged in Directive 91/440 and foreshadows the Commission's proposals by including national services. This is the case in Germany, where free access for freight has been established (but on a reciprocal basis in the case of foreign operators), the United Kingdom, where there is also free access for freight, and the Netherlands where free access also covers passenger services. Despite these provisions, new competition is emerging only with difficulty. The level of charges and the way in which they are calculated have no doubt discouraged newcomers in Germany, who remain confined to a few freight operators.

Although for the time being access is free in the Netherlands, only a few freight operators have arrived on the scene together with one passenger operator ("Lovers Rail") controlled by French interests; in the United Kingdom, only two freight operators have emerged, one of which was subsequently taken over by the main operator. Attempts can be made to explain this situation without referring to the reticence or even obstruction of infrastructure managers vis-à-vis newcomers in respect of paths and pricing offers: it has to be admitted that such negotiations and the technical identification of paths available are extremely complex.

This type of access right also exists, and has existed for some time, at the national level in the United States where Amtrak only owns 450 miles of track itself but has access to a further 24,000 miles of the American network owned by private freight companies on payment of a fee. Access rights also exist in the case of private freight companies and the Surface Transportation Board, which is the successor to the Interstate Commerce Commission, in accordance with competition policy, requires companies owning track to allow access by other companies. This is also true in Japan, where the only freight company has access to the whole regional company network subject to payment of a fee following privatization of Japan National Railways. In New Zealand, there are access rights for sections that carry less than a certain tonnage or specified number of passengers, but in this instance the railway company remains vertically integrated (infrastructure plus operation).

It is significant to note that this concept of access rights does not appear at all in the schedules of commitments even though it could be the technical vehicle for multilateral liberalization. If this should be the case in the future, non-discriminatory allocation of rights would raise problems similar to those for allocating air transport slots (whether or not this is the government's responsibility, allocation principles, etc.) and to a lesser extent interconnection and the allocation of frequencies in telecommunications. The question of non-violation (access denied because no path is available) would be a particular problem.

In parallel with this movement to separate the infrastructure from operations and to allow marginal opening up of access rights, more radical privatization initiatives have been tried out in developed countries, as well as in countries with economies in transition and developing countries under the auspices of the World Bank.

Up until now, the United Kingdom has gone the furthest along this road. As this experience in some respects constitutes a form of international liberalization of rail transport, it needs to be described in some detail. In the United Kingdom, the principle of separating the two activities has been taken to extremes: the infrastructure has been given to a company, Railtrack, which was privatized in May 1996; passenger rolling stock was divided up among three companies subsequently sold to the private sector which lease it to the operators. The system was intended to reduce the entry costs for concession-holders. Passenger services were handed over to 25 companies, some of them controlled by foreign interests, in the form of concessions for periods of seven to 15 years. Track renewal and maintenance was also given to companies sold to the private sector, which compete in order to win contracts.

Two government bodies, the Office of Passenger Rail Franchising (OPRAF) and the Office of the Rail Regulator (ORR) grant the concessions, regulate prices and the terms of access to the tracks, specify minimum service levels and, for certain categories of fare, maximum fares. In order to protect the concession holder, open access to passenger services is limited to routes where there is no service or which account for a very small part of the concession holder's income. There is open access for freight, as mentioned above, but with even more limited results.

The immediate effect of privatization was a sharp rise in subsidies given to operators because they had to meet costs not faced by integrated operators: the charges for using the infrastructure and rolling stock (grants to the British Railways Board: 1993/4: £1,121 million; 1994/1995: £1,984 million. Subsidies to concession holders, 1996/1997: £2,090 million; estimated subsidy for 2003: £1,169 million.) Operators have nevertheless undertaken to reduce the subsidies by half within seven years and some routes could even become profitable. It is as yet too early to draw any lessons from this experience. Some competition can be seen on certain routes, the level of services has improved and operators have started to renew the stock, but the OPRAF is still critical of the level of services provided by certain concession holders and Railtrack has not reached the investment objectives fixed.

As part of its loan activities for restructuring railways, the World Bank, after having for a long time promoted autonomous operators rather than the government (in Pakistan, Colombia, Korea, Senegal, Mali, Yugoslavia) now encourages the granting of concessions. This was the case in Argentina, Côte d'Ivoire and Burkina Faso.

Concessions obey the logic of government procurement rather than that of market access within the framework of the GATS. It should nevertheless be noted that no member of the Plurilateral Agreement on Government Procurement has made any commitment on rail transport services.

The relative importance of different modes of supply and obstacles to trade are fairly closely linked to the structure of railway companies and their evolution as described above. In the case of mode 1, for a long time international transport of passengers and freight only consisted of joining successive national segments from the point of view of both fares and the technical and legal responsibility for transport. No single entity was responsible for an international journey, the freight or the passenger being passed on from one monopolistic network to another. In mode 1, therefore, there was no competition, except in the case of transit between the same two points using different routes (Rotterdam-Genoa through Germany and Switzerland or through Belgium and France). It was only following the arrival of high-speed trains, on the one hand, and freight corridors, either freightways or freeways, on the other, that basic commercial concepts emerged; for instance: the one stop shop, harmonized commercial policy, the pooling of revenue or even more simply common accounting methods to allow the profitability of an international transport operation to be assessed. In spite of the noteworthy efforts made over the past 40 years, national operators still face many technical obstacles when operating beyond their own borders, even when such operations are allowed and they have an ad hoc path: the type of electric power, different gauges, signalling systems, braking systems, commercial speed limits, height of railway wagons, axle load and technical standards for wagons.

With regard to mode 2, there does not appear to be any restrictive legislation anywhere. On the contrary, there is cross-border cooperation among railway companies to attract certain customers and incite them to use rail transport services in mode 2 (young peoples' rail passes, Eurorail cards, for example).

There was no trade under mode 3 as long as rail transport remained a government monopoly. The establishment and development of private companies (provided that a majority shareholding by a foreign company is allowed), together with the gradual introduction of access rights, or even eventually cabotage rights, subject to establishment in the country concerned, now make rail transport under mode 3 increasingly possible. One example of this type of trade is the recent purchase of minority shares in the largest Mexican railway by several American railway companies and the planned purchase of minority shares in another Mexican network by another American railway company. In both cases, however, these are minority shareholdings. The increase in concessions, which are often given to foreign groups that then have to become established in the country concerned, also allows the development of trade under mode 3, but in this particular case the concession holder is given a monopoly, at least a regional monopoly, which restricts access by third parties and the issue concerns government procurement rather than market access. As far as mode 4 is concerned, there is always a marginal flow of technicians and engineers, particularly to developing countries. The increased number of concessions has seen this flow broadened to include managers.

Following this brief description of the economic aspects and before analysing in detail the commitments undertaken, some comments on the MTN.GNS/W/120-CPC classification for this sector are called for. A study of this sector highlights what appear to be a certain number of instances in which the classification is not adapted to the true situation. Moreover, the classification itself seems to be ambiguous in some cases.

PRELIMINARY REMARKS ON THE RELEVANCE OF THE CLASSIFICATION TO COMMITMENTS

Rail transport services have been divided into five subcategories, namely, passenger transportation, freight transportation, pushing and towing services, maintenance and repair of rail transport equipment and supporting services (terminal services, cargo handling services, other support services).

In this structure, passenger transportation by railway (11.E.a) includes both interurban transportation and urban and suburban transportation, that is to say underground or elevated railways, whereas in economic and regulatory terms it can be argued that urban transportation by railway has characteristics that are closer to those of competing means of transport such as motor buses, tramways and trolley-buses: A concession regime with public service obligations regarding networks and schedules compensated by operating subsidies and public financing of investment, often State-owned, managed by a common transit authority. Additionally, as mentioned in detail in paragraph 24 of document S/C/W/60, it appears that for light rail vehicle (LRV) networks the distinction between rail transport and urban road transport is blurred and even disputable. In practice, of the ten Members that have made commitments on rail transport services, only one has distinguished underground from other passenger transport and introduced similar qualifications to those of the other urban transport systems.

The classification singles out an activity, "pushing and towing services", which is generally not performed by an entity separate from the one operating the transport (except sometimes in the case of the private connections where these services are generally supplied on their own account and therefore escape the discipline in the Agreement). Five Members have, however, offered these services, three of them without limitations and two of them with limitations (joint venture in one case, attribution of a concession in the other case). Technically speaking, the exercise of this activity by a separate entity from the one(s) operating the transport and the network implies access to the network and some kind of coordination with the operator(s) of the rail transport. If access to the network was to be denied or made subject to the payment of a large fee or coordination with the operator was refused, there might be a case for an action for non-violation and for invoking Article VIII. The question therefore arises whether or not it is necessary to establish more precise rules to guarantee access to the network and safeguard competition.

It should be noted that that maintenance and repair activities are limited to the transport equipment and not the maintenance and repair of the railway infrastructure, which is covered by the construction items of the CPC (CPC 51310 "construction work of highways, streets, roads, ... railways and airfield runways", CPC 51320 "construction work of bridges, ... elevated highways, ... tunnels, and subways").

"Supporting services" are not described in great detail: "terminal services, except cargo handling, and other supporting services for railway transport" whereas similar headings for other modes of transport are much more detailed in the classification. This imprecision may be one of the explanations for the very low level of commitment in this subsector (four countries) and for the need felt by one country to have recourse to sui generis concepts. It might be worthwhile envisaging the development of a more detailed list based on the concepts of the industry. Finally, Members will find a detailed assessment of the changes suggested for railway transport in the CPC/Rev.1 classification in document S/CSC/W/6/Add.5, dated 4 June 1997, paragraph 5, and a summary of those changes and of their potential impact on existing commitments in document S/CSC/W/9, dated 9 October 1997, paragraphs 31 and 32.

ANALYSIS OF COMMITMENTS

The commitments at subsectoral levels (18 out of 47) concern repair and maintenance, a subsector which has been offered by 18 countries out of 21 having made commitments in the sector. For repair and maintenance, the regime offered appears extremely liberal as full market access in consumption abroad has been granted in 16 cases out of 18 and in commercial presence in 12 cases, while partial restrictions were scheduled in two cases (joint venture, space and capacity constraints). It is worth noting that 12 countries have considered the supply of these services in cross-border mode as not technically feasible, while five others deemed it feasible. It may be worth considering a harmonized and refined solution to this question of technical feasibility, taking into account the progress in electronic tele-maintenance, which now makes certain operations technically possible.

Passenger transportation and freight transportation have each been offered by ten countries, eight of which have offered both subsectors. In two instances, the scope of the sector offered has been limited: exclusion of high-speed trains in one case (United States) and of bulk liquids, gases and mail in another (Brazil). Market access has been opened up in four cases out of ten in model 1, while this mode was considered technically infeasible for passengers by two countries and for freight by one country. This divergence is related to the legal qualification of international trains. A harmonized solution to this question may be worth studying.

Such a solution could take into account the general commercial evolution in both high-speed passenger and freight trains, moving from a regime where a journey meant joining together national segments in terms of pricing, management, and revenue-sharing towards a more cooperative and integrated regime among railway companies (one-stop shop, common exploitation, transnational freight alliances, etc.). Here again, the consumption abroad regime is extremely liberal (ten out of ten cases and nine out of ten cases respectively). Market access to the key commercial segment, commercial presence, has been fully offered by two Members, one of them having however kept an unbound entry for national treatment restrictions.

For passengers and freight, the other seven and six countries respectively made partial commitments on commercial presence, but then introduced requirements on concessions authorizations, incorporation, joint ventures and investment limitations, and, with regard to national treatment, conditions relating to the nationality of executives, together with additional concession requirements. Pushing and towing services have been offered in total by five Members with no limitations whatsoever in mode 2. Three countries also have a fully liberal regime for modes 1 and 3, while two countries imposed restrictions under mode 3 (joint ventures and concessions). No specific national treatment limitations were entered. Supporting services for rail transport have been offered by four countries (one of them limiting the scope to security and cleaning), with a full liberal regime on consumption abroad. Two of these countries introduced limitations on commercial presence (joint ventures and horizontal limitations). Finally, regarding mode 4, all the subsectors of rail transport seem to adhere to the classic norm of "unbound except as indicated in horizontal commitments".

In total, it appears that the level of commitments in rail transport is extremely low (13 per cent of the Members for maintenance, 7 per cent for passenger and freight transportation, with many limitations). This can certainly be explained by the existence of national monopolies. However, three of the activities (maintenance, support services, and pushing and towing) can technically and economically operate outside or inside the classic monopoly scope. There is certainly room for commitments under modes 1, 2 and 4, whose de facto regime seems in general relatively liberal. It is also noteworthy that the various regulatory reforms separating infrastructure and transport, privatizing, granting concessions and establishing competition did not translate at all into commitments undertaken. Their binding in WTO schedules may require additional work in the future on scheduling (additional commitments on access to/use of the network), monopolies, competition and the government procurement regime (for the concessions).

As far as MFN exemptions are concerned and setting aside the "all sectors derogations" it appears that two types of exemption affect the supply of rail transport services: those concerning land transport in general and those concerning rail transport in particular. In seven cases out of ten, the general land transportation exemptions concern regional agreements in South and Central America. The three other cases concern reciprocity requirements, among which one case of tax reciprocity (VAT). There are five proper rail transportation exemptions: three of them have been made in identical terms by Central European countries to cover existing or future agreements regulating traffic rights and operating conditions, one concerns preferential treatment for leasing charges for railway wagons for specified neighbouring countries, and one a reciprocal deduction of tax on earnings from the use of rolling stock. These MFN exemptions have a noticeable effect on the commitments as seven Members out of the 15 that have made exemptions have made commitments (two out of the ten have made general land transportation exemptions, all of the five have made proper rail transport exemptions).

8

Roads for Development

INTRODUCTION

The proposed Roads for Development program (R4D) will be designed for 5 years starting in 2012. Subject to approval to proceed by an AusAID concept peer review, the program will be jointly designed by the ILO and AusAID. Subject to final approval by AusAID and the ILO, the program will be implemented by the ILO with AusAID funding and ongoing policy engagement. R4D is expected to become the main donor program for improving and maintaining rural roads in Timor-Leste in the coming years with an explicit objective of strengthening institutional capacity and developing systems and procedures for the management of rural roads.

This Concept Note was prepared by the ILO and is the outcome of a wide range of consultations that have taken place since mid-2009 between government stakeholders, AusAID and other donors involved in rural roads development, culminating in a stakeholders' workshop on 3 March 2011. The workshop was attended by representatives of the Ministry of Infrastructure (MoI), Ministry of State Administration, Ministry of Economy and Development and the Secretary of State for Vocational Training and Employment (SEFOPE). Leading donors in the roads sector also attended the workshop, including AusAID, Asian Development Bank (ADB), European Commission (EC), World Bank, JICA and Norway. In this workshop and during various consultative meetings, the Government of Timor-Leste (GoTL) has indicated that it supports the proposed program as rural roads development is one of its key priorities.

The concept for the proposed program builds on the lessons learned from completed and on-going rural roads and capacity development programs and projects in Timor-Leste, in particular the TIM-Works Project which is being implemented by the ILO through SEFOPE. TIM-Works is currently the largest program in the rural roads sub-sector. TIM-Works started in 2008 and is co-funded by the EC, Norway, Ireland, AusAID and GoTL. The ILO's effectiveness in implementing TIM-Works was confirmed by the findings of the independent mid-term evaluation and the results oriented monitoring mission engaged by the EC. Given its experience, technical expertise and demonstrated performance in labour-based rural roads development and capacity building projects in Timor-Leste and internationally, the ILO is the most appropriate implementing partner for R4D.

There is broad recognition among stakeholders of the need to ensure and sustain adequate rural roads access as a pre-requisite for local economic development and poverty alleviation and to provide rural employment opportunities. This recognition, together with the results achieved through labour-based rural roads works, provides an opportunity to up-scale and institutionalize the successfully demonstrated activities in this sub-sector. The current proposal reflects priorities of GoTL and the Australian Government. The main thrust of the proposed R4D is to develop and institutionalize adequate capacities and instruments in the public sector that will enable GoTL to effectively and equitably plan, budget and implement investments in rural road rehabilitation and maintenance.

R4D will adopt a holistic approach and will build synergies and complementarities with other on-going and planned interventions in the rural roads sector. Through R4D, AusAID and the ILO aim to take a leading role in policy dialogue with the government for the sub-sector, particularly to positively influence budget allocations. The program will support government policy formulation and help develop strategies and operational procedures for the management of rural roads. It will also take a leading role in coordinating donors in the sub-sector.

Considering current operational capacity and funding constraints within GoTL, it is envisaged that, complimentary to its capacity building activities, R4D will initially provide direct implementation support and investments in rural roads. R4D will directly manage funding of rural road works from donor resources and support the government to implement rural road works from its own budget. As capacities within GoTL develop, it is foreseen that R4D's direct investments and operational support will be gradually phased out, with the GoTL increasingly taking over these responsibilities. As capacity increases Australia will also look to provide funding to the rural roads sector through GoTL financial management systems to reduce transaction costs and to support policy dialogue on budget allocations and systems strengthening.

Whilst the proposed R4D will build on the lessons from TIM-Works and other similar interventions, R4D will have infrastructure as the central objective with employment generation being a secondary but still important aspect. Given its focus on skills development and employment generation, SEFOPE was an appropriate institutional home for TIM-Works. However, with this shift in emphasis it is deemed appropriate and more sustainable to place R4D with MoI as the line ministry that is responsible for the rural roads network. R4D's capacity building activities will focus on assisting MoI in establishing its capacities at the national, regional and district levels. This will be done in a phased manner, taking into account absorption capacities and the necessary alignment with the on-going decentralization process. Considering the current uncertainties regarding the pace and direction of the decentralization process, the design of R4D will incorporate sufficient flexibility to ensure the continued effectiveness of the capacity building activities throughout the life time of the program.

CONTEXTUAL ANALYSIS

KEY DEVELOPMENT ISSUES

Timor-Leste is one of the least developed countries in the region. Its non-oil economy is characterised by slow and volatile growth, although there has been recent strong growth in urban centres, Dili in particular. The economy is essentially agriculture-based, with about 72 per cent of the total population of 1.1 million living in rural areas. Subsistence farming provides the main source of livelihoods for the large majority of rural people with limited production of agricultural products, mostly coffee, for cash income. Agricultural productivity is low compared with other countries in its region. Timor-Leste is off track to achieve most of the Millennium Development Goals.

The country has experienced conflict and instability. At independence in 1999 more than two thirds of its infrastructure was destroyed and much of its administrative capacity returned to Indonesia. Like all post-conflict fragile states, Timor-Leste has a high level of risk of returning to conflict which may again prevent development and promote further social unrest. However, since 2008 Timor-Leste has experienced a period of stability and as a result GoTL has now explicitly shifted its focus from stabilisation to what it terms "growth and development for all".

Despite development efforts in recent years, the poor condition of basic infrastructure in general, and the road network in particular, remains a key constraint. In a transport sector study commissioned by the ADB in 2006 the poor condition of the roads was cited by the interviewed households as the major cause of poverty and the largest constraint to local economic development. It was followed by the lack of access to income generating opportunities, which is closely associated with the existing poor road conditions that prevent easy access to market activity centres. Another serious problem for the economy is the inability to create employment opportunities for the rapidly expanding labour force, particularly in rural areas. Due to the lack of rural employment opportunities, Timor-Leste is experiencing large-scale rural-urban migration. This immigration results in strong pressures on scarce resources, facilities and services in overstretched urban areas, particularly in Dili. The limited job prospects in Dili, have furthermore promoted the growth of a sub-culture of young men engaging in organized crime and violence, with the potential to spill over into civil unrest.

THE CONDITION OF THE RURAL ROADS NETWORK

Timor-Leste has a relatively dense road network, with a total estimated length of 6,036 kilometres. It includes 1,426 km of national roads, 869 km of district roads, 716 km of urban roads and 3,025 km of unpaved rural roads.

Overall, the road network is in a bad condition. A recently completed study funded by the EC surveyed 730 km of rural roads accessible by vehicle and found about 69 per cent of these to be in a bad or very bad condition. It is therefore estimated that 90 per cent of the total rural roads network is in a bad or very bad condition. National and district roads are also in poor condition. Systematic routine, recurrent or periodic maintenance on maintainable roads hardly takes place (with the exception of maintenance activities on some donor-funded projects like TIM-Works).

Roads constitute the primary mode of transport in Timor-Leste, carrying 70 per cent of freight and 90 per cent of passengers. Due to the poor condition of rural roads, rural people face increased travel times and transportation costs and are isolated in terms of access to social and economic facilities and services (like local markets, schools, health facilities, job opportunities, government services and banking services). The current rural road access problems seriously constrain the scope and incentives for local economic development and agricultural diversification.

The Complexity of Procurement and Implementation of Rural Road Works

The management of the rural roads sector is fragmented with different government agencies and various procurement processes involved. GoTL significantly increased its capital budget in 2011. While the capital development allocation from the consolidated fund of US$89 million was less than in 2010, the budget included a new and substantial Infrastructure Fund with an allocation of US$599 million in 2011 that is projected to increase to US$672 million in 2015. Three quarters (US$ 449 million) of the Infrastructure Fund in 2011 is allocated to national electricity generation and transmission infrastructure. The Infrastructure Fund includes a substantial allocation to the MDG Suco program in 2011 (US$65 million) to provide houses across the nation along with small-scale infrastructure in solar power, water and sanitation and roads in addition to other community-based works. The consolidated fund includes a substantial allocation to the Decentralized Development Program (PDD) (US$44 million) for road, power, and water and sanitation infrastructure at the aldeia (sub-village), suco (village), sub-district and district levels. Programs such as the PDD and the MDG Suco Program are meant to promote more equitable development of rural areas while giving ownership to people to rebuild their communities.

The procurement of works under the MDG Suco Program and the PDD will be managed by the Ministry of State Administration for works up to a value of US$150,000, whereas the newly established National Development Authority (NDA) will manage procurement of works under the PDD for works from US$150,000 up to US$500,000. NDA will also have a major function as a monitoring and certifying institution for all works planned and implemented under the Infrastructure Fund.

Line ministries (including MoI) will continue their roles as previously and can procure works between US$500,000 and one million dollars whilst larger works will go through the Major Projects Secretariat within the Ministry of Finance and the Procurement Commission under the Office of the Prime Minister.

Although not managing the procurement processes under the PDD, MoI still plays a key role in support to planning, contracts preparation and contracts supervision for works under the PDD. It seems likely that MoI through its Directorate for Roads Bridges and Flood Control (DRBFC) under the Secretary of State for Public Works will continue to play a key role in rural roads development particularly with regards to planning, design and supervision of works. A Rural Roads Department under the Director of DRBFC is being established and as such likely to become the hub for rural roads planning and management, even though it will not be directly involved in the procurement of works. While support for road works has increased in recent years, the priority of the government has been national and district paved roads. Despite the development need, funding for rural roads has been variable. The 2011 Budget has no provision for rehabilitation and maintenance of rural roads.

Capacity Constraints in Delivering Investments in Rural Roads Development

Current technical and managerial capacity constraints in the public and private construction sector in Timor-Leste seriously inhibit GoTL's capacity to manage and implement investments in infrastructure, including investments for rehabilitating and maintaining the rural roads network (using small-scale domestic contractors). Due to a lack of professional staff in the government's technical agencies, combined with a limited capacity in the private construction industry, the delivery rate of public works schemes has in the past been very low. With specially managed infrastructure funding starting GoTL's 'Pakote Referendum Package' in 2009 and continuing through PDD and the MDG Suco program, disbursement levels have increased significantly, however without matching planning or supervision capacity resulting in inadequate quality and inefficiency.

DRBFC is responsible for the maintenance and operation of all roads in Timor-Leste. At the central level, DRBFC has sections for Planning and Design, Operations, and Administration and Finance. Procurement is co-ordinated and administered by the Procurement Unit within the MoI. Each of the five regional offices is headed by a regional engineer, with roughly a supervisors and an assistant supervisor for each district.

Although there has been a massive increase in budget and work commitments of DRBFC, the number of engineers has remained static since 2001. DRBFC uses the private sector as much as possible for design, construction and contract supervision as well as construction itself to reduce the work load of its staff. While this reduces the workload for the line staff, it does not remove the need for good management and supervision skills. At the same time capacity constraints of the domestic private construction sector limit the scope for outsourcing work to the private sector.

Although growing in numbers, the number of qualified domestic contractors and engineering consultants in Timor-Leste is still too low, especially at district level, to meet implementation requirements for investments in the rural roads network. Many new small contractors have appeared at the local level due to recent government contract packages but they are lacking in technical skills, as well as the necessary experience and discipline of effective contract management.

GoTL is at present preparing the development of a register of contractors in a number of classes according to their technical and financial capabilities. This register, once established, could act as a prequalification list for different-sized construction contracts. All contractors need to be registered with the contractors' association (AECCOP). AECCOP was responsible for issuing contracts to its members under the GoTL's small infrastructure procurement process in 2009.

Timor-Leste Government Priorities, Policies and Strategies

To substantially improve the quality of life of its people and to reduce the incidence of poverty, GoTL has committed to the development and improvement of a well-connected and coherent road network and other key infrastructure, which are seen as being fundamental to the country's development. The need to promote the development of the currently nascent domestic private sector is also recognized and prioritized as private sector capacities are much needed, for example to deliver investments in road infrastructure.

The improvement of rural infrastructure and the promotion of employment opportunities are hence key development priorities for GoTL. Infrastructure has been the official National Priority No 1 from 2009 to 2011, which includes rehabilitation and maintenance of rural roads and the creation of employment through promotion of labour-based methods in public works. The elevated status of infrastructure as the number one priority has also been mirrored in the increased GoTL budget allocation to infrastructure over the last couple of years. However, rural roads have not been prioritized within these substantial allocations to infrastructure.

Public infrastructure works programs that provide essential basic infrastructure, generate short term employment, and contribute to the development of long-term and sustainable job creation also constitute pillars of the 2011 National Priority Working Group on Rural Development. The Ministry of Economy and Development, the Secretariat of State for Vocational Training and Employment and the Ministry of Infrastructure participate in this working group.

The summary of the Strategic Development Plan (SDP) for the coming 20 years emphasises infrastructure, including roads, in achieving accelerated sustainable development. The SDP acknowledges that scaling up GoTL's fiscal envelope is not enough to deliver improved infrastructure and that there is a need to increase the investment in the building of a national capacity to implement the SDP. A number of other current and past strategies have also defined roads as a priority but have not been officially adopted or utilised and will be superseded by the SDP. The proliferation of planning documents with limited government ownership highlights the need for practical engagement that will strengthen GoTL's ability to manage its resources for rural roads. R4D will be fully embedded in GoTL structures which will facilitate GOTL-led planning and implementation of rural roads works.

Priorities of the Australian Government and the UN

Improving infrastructure, including through labour-intensive initiatives, is a key objective in the Australia Timor-Leste Country Strategy 2009 – 2014. A key priority for the Australian Government is building local capacities so that Timorese professionals can work on Timorese challenges. The strategy emphasizes the need for strong donor coordination to help ensuring real progress on national priorities and to avoid the creation of unsupportable financial obligations for GoTL. Australia aims at playing a stronger role in donor coordination, policy analysis and policy dialogue.

R4D is well aligned with outcome 2 of the United Nations Development Assistance Framework: By 2013, vulnerable groups experience a significant improvement in sustainable livelihoods, poverty reduction and disaster risk management within an overarching crisis prevention and recovery context. The program also contributes to ILO's Decent Work country outcomes relating to increased rural employment and improved safety nets through infrastructure investment, livelihood improvement and local business development.

Related Donor-supported Programs and Projects

Road sector development in Timor-Leste relies heavily on external assistance. Support mainly focuses on up-keeping the main roads with few projects focussing on the improvement and maintenance of the rural roads network and related capacity building activities.

ADB: Early ADB support included two emergency infrastructure rehabilitation projects and the Road Sector Improvement Project. Together with AusAID, ADB is financing the US$15 million Infrastructure Management Project. Through its Road Network Development Sector Project (RNDSP), the ADB is providing a US$46 million grant that includes the improvement of 232 km of national roads and the establishment of a maintenance program for national roads. The ADB's Japanese-funded, US$3 million, village access project Our Roads, Our Future will support local governance and community-based infrastructure works. This includes the improvement of selected village roads in three districts and enhancing the involvement of communities in road works and small community infrastructure. The ADB is also considering supporting two GoTL national roads projects – the Road Network Upgrading Project on the north coast and the Road Network Rehabilitation Project in the west of the country. The first of these projects would be supported by loans from the ADB and Japan.

Japan: The Government of Japan has provided more than US$30 million in grants for both technical assistance as well as capital works (focussing on national and district roads). In addition, through the Japanese Fund for Poverty Reduction, US$3 million has been allocated for Our Roads, Our Future. Japan is considering a loan to support the Road Network Upgrading Project through the ADB.

World Bank: GoTL has formally applied for a US$20 million grant to implement the Bank's Road Climate Resilience Project. The project will invest in the Dili-Ainaro road to improve its climate resilience. It will include maintenance funding linked to the ADB RNDSP, pilot testing of emergency planning and response system, and training for MoI staff and contractors. The four-year project is expected to commence in July 2011.

European Commission: For a number of years the EC has provided some US$ 50 million through Rural Development Programs (RDP I, II and III). As part of these programs, road rehabilitation works were funded. The EU also co-finances TIM-Works. Currently the EU is preparing RDP IV. Rural road rehabilitation and capacity building of small-scale contractors is one of the components of RDP IV (*i.e.* the € 10 million CORE project with MoI). CORE is a 4 year project planned to start in mid 2011. CORE will complement R4D by building private sector capacity in contract management and delivery. CORE capacity building will target 64 local contractors for rural road rehabilitation and maintenance. CORE will have close links to MoI and the contractors trained by CORE will be used by R4D to implement rural road maintenance and rehabilitation. CORE will be implemented by the ILO which will facilitate close coordination.

AusAID: AusAID has contributed A$2.25 million to TIM-Works. AusAID is co-financing the ADB's Infrastructure Management Project. AusAID is also funding the ILO-implemented Youth Employment Promotion Project (YEPP), which includes a component for creating employment opportunities for youth in the maintenance of rural roads. SEFOPE will continue implementing labour-intensive employment schemes as a social safety net but the AusAID support to rural road infrastructure will in future be channelled through the new R4D.

Norway and Ireland: Norway and Ireland have provided co-financing support to TIM-Works.

Consultations have been undertaken with the key donors in the rural roads sector. Arrangements for coordination and collaboration through R4D have been discussed and there is general support for the proposal and a willingness to cooperate. The ADB has indicated that they support AusAID and the ILO taking a lead role in the rural roads sub-sector and will coordinate with R4D.

ROADS FOR DEVELOPMENT

OUTLINE OF THE CONCEPT

R4D will support GoTL in planning, budgeting and implementing investments in rural road rehabilitation and maintenance, using labour-based methods where appropriate. While the ADB is the lead donor on national roads, the proposed R4D with AusAID support will lead on rural roads. Accelerating capacity building and increasing investment levels for the rehabilitation and maintenance of the rural roads network of Timor-Leste are considered key priorities for GoTL and AusAID. As capacity building is a long-term process, the proposed support through R4D should reflect this and it is proposed that R4D be designed for a period of 5 years.

Capacity building will be central to R4D from inception and will mainly focus on strengthening capacities in the public sector by supporting MoI in establishing functional management and technical capacities as well as developing supporting policy/strategy and operational tools. AusAID has developed a staged approach to capacity development that will be integrated into the design of R4D. The approach sets out four stages of capacity for different functions – dependent, guided, assisted and independent. The design will consider the stages that MoI (and district administrations) are at and develop appropriate capacity building activities for these stages. Subject to confirmation by the design, achieving an 'assisted' level of capacity is considered a reasonable end-of-program outcome.

The design will specify requirements for a detailed needs analysis at the inception of the program to guide capacity building activities. Capacity building activities will be guided by GoTL's Technical Assistance Policy which is under development. Preliminary discussions on this policy have emphasised the need to distinguish between different technical assistance functions, namely strategic advice, operational work and capacity building. R4D will initially also provide substantial direct operational assistance and investment support in the planning and delivery of rural road rehabilitation and maintenance. This direct support will be implemented in partnership with the GoTL agencies, in particular with the Department of Rural Roads within MoI. A mixed approach including direct implementation is appropriate given the current capacity constraints in the public sector and the clear and pressing need for the rehabilitation and maintenance of rural roads. To focus solely on capacity building at this stage would likely result in a further deterioration in the rural road network before adequate public sector capacity is built to manage the necessary investments. Direct support will also demonstrate the process and benefits of rural road rehabilitation and maintenance and need for appropriate budget allocations to this sub-sector.

Direct implementation will complement capacity building by providing an effective platform for on-the-job learning for the professional staff of MoI. However, care will need to be taken to ensure that direct implementation is supporting, rather than substituting for, capacity building activities. Co-sponsoring (jointly funding) with GoTL for the placement of district-level staff is envisaged to ensure core operational capacities at district level. As capacities are being developed and assuming GoTL investment levels and staffing increase, R4D's role will gradually evolve and focus increasingly on advisory support.

R4D aims to integrate labour-based methods, using a balanced mix of labour and equipment, into the rural roads system where cost effective and appropriate. Integrating labour-based methods in the sub-sector will reduce the need for imported equipment and materials and instead increase the use of local resources to implement road rehabilitation and maintenance works. Labour-based road works will provide short-term employment opportunities and an injection of cash income into rural communities. Wage rates will be set in consultation with GoTL at a level that ensures an adequate supply of labour without distorting local labour markets. Contracting out road works will contribute to private sector development. Contractors trained under CORE and TIM-Works will be eligible to bid for works under R4D.

R4D builds on experiences of earlier rural roads development projects, in particular TIM-Works and other relevant capacity building projects (like the AusAID-funded Rural Water Supply and Sanitation Program). Lessons learned from TIM-Works, as well as similar programs elsewhere in the world, indicate the appropriateness, cost-effectiveness, feasibility and replicability of:

- The application of labour-based approaches
- The quality control and quality assurance methods used in these programs
- Capacity building methods that include a strong emphasis on on-the-job training and coaching, supplemented with class-room training

- The targeting of beneficiaries in the creation of short-term employment
- The use of local small-scale private contractors and community contractors for rehabilitation and maintenance works respectively.

R4D will have national coverage. Selection of specific road works will be undertaken in accordance with GoTL policy but R4D will promote equitable national coverage in line with needs, taking into consideration past, on-going and other planned investments in the sector, including linkages to investments in national and district roads.

R4D will take a leading role in supporting GoTL planning and implementation of rural road works and coordinate activities with other donors. At present, donor-supported activities at policy and operational level in the rural roads sector are not well coordinated. However, with R4D providing the necessary advisory support to GoTL leadership of rural roads, including a central role for GoTL in donor coordination, it is expected that the overall effectiveness and sustainability of the interventions of the various stakeholders in the rural roads sector can be improved.

R4D's capacity building activities will be aligned with GoTL's decentralization process. Considering the uncertainty of the direction and pace of this process, R4D's design will be sufficiently flexible to ensure that its interventions can respond to the actual implementation of decentralization. The design of R4D will need to specify criteria to guide a flexible approach to decentralization to avoid too much flexibility leading to drift in the program's objectives and priorities.

Despite the uncertainty around decentralization, it is logical that the central counterpart institution for R4D is the new Rural Roads Department within the DRBFC of MoI. MoI should be in a position to technically oversee and influence rural road rehabilitation and maintenance. However, the R4D will clearly have to work with all the processes and institutions involved in rural road management to be effective. R4D will operate in a dynamic environment so there needs to be flexibility built into the design for necessary adjustments throughout the duration of the program.

As well as a lack of clarity around responsibility for rural roads, there is currently no national plan or specific budget allocation for rural roads. R4D will support the MoI in developing plans and budget requests. It is expected that with increasing management capacity, annual plans and sound systems, budget allocations will increase and eventually reach a sustainable level for the ongoing management of the rural road network. AusAID and the ILO will pursue policy dialogue with MoI, the Ministry of Finance, other relevant agencies and relevant ministers to build support for increased budget allocations to rural roads.

R4D activities will be closely coordinated with other relevant capacity building projects (including CORE and Japanese and ADB-funded capacity building projects). Technical assistance under R4D will include support to DRBFC's regional offices and district administrations involved in rural roads development and maintenance. Coordination with the AusAID-funded YEPP will focus on establishing linkages related to the provision of sustainable employment opportunities for youth in the domestic construction sector, through the employment mediation services of the Employment and Career Guidance Centres network established in partnership with SEFOPE.

Goal and Objective

The overall goal of R4D is to contribute to rural poverty reduction and livelihood improvement by supporting the Government of Timor-Leste's objective of a rehabilitated, upgraded and maintained rural road network. The immediate objective of R4D is: The Government of Timor-Leste effectively and equitably plans, budgets and implements investments in rural road rehabilitation and maintenance, using labour-based methods where appropriate.

Budget Estimate

The envisaged AusAID contribution to R4D is up to US$30 million. These contributions are subject to clarification during the design of R4D and annual budget appropriations. As capacity building is a long-term process, further funding may be required beyond the currently envisaged 5-year implementation horizon. It is expected that GoTL will contribute to the cost of national staff working with R4D and increase its own budget for rural road investments during the life of R4D.

The direct rehabilitation works will cover 600 km of rural roads and will include spot improvement of problem sections and in other cases rehabilitation of the entire road. Wherever possible, highly technical engineering solutions will be avoided. For budgeting and programming purposes the direct investment costs for rehabilitation works have been estimated at an average of US$30,000 per kilometre. R4D will directly implement maintenance of 750 km of roads. The cost of routine maintenance for roads with no maintenance backlog is estimated at an average cost of US$800 per kilometre per year. The design process will test the appropriateness of these average cost estimates. R4D will support MoI to prepare detailed cost estimates for rural road rehabilitation and maintenance in different geographic areas and integrate these estimates into planning and budgeting. The preparation of cost estimates will be informed by the experience of TIM-Works and integrated into a unit cost database that was developed in MoI with ADB support.

Delivery and Implementation Arrangements

To enable an effective and smooth transition from the on-going TIM-Works, it is recommended to retain to the extent possible technical assistance personnel and trained technical staff of the TIM-Works project, whilst building the R4D structure within MoI. This will require TIM-Works to be extended from its current completion date of 30 June 2011 to at least the start date of R4D, anticipated to be February 2012. Technical responsibility for R4D will be assigned to a Project Coordinator, with technical backstopping support provided by the ILO Regional Office in Bangkok. Day to day administrative and financial services support will continue to be provided by the ILO Office for Timor-Leste and Indonesia in Jakarta, as delegated to the ILO Liaison Officer in Dili. While the majority of the staff will be national and recruited by MoI, international staff and consultancy inputs will also be required, considering current capacity constraints in Timor-Leste. By ensuring that GoTL counterpart staff are available, effective information and knowledge exchange can take place. Locally-recruited technicians and engineers will be employed under GoTL remuneration conditions and initially paid for by R4D but increasingly by GoTL (this practice is successfully applied in TIM-Works and YEPP).

Gender Issues

R4D will be designed in line with ILO policies and processes on gender equality and AusAID guidance on integrating gender equality into activity design. The ILO's mandate on gender equality is to promote equality between all women and men in the world of work. This mandate is grounded in the International Labour Conventions. The ILO's gender policy states that mutually reinforcing action to promote gender equality should take place in staffing, substance and structure. The policy is being made operational through an ILO action plan on gender equality, which facilitates gender responsive design, resourcing and delivery of the work of ILO. Further, the employment intensive investment branch of the ILO work at several levels to promote decent job opportunities for women in construction. The Guide on Labour Policies and Practices recommends several practical steps for achieving a gender mix on site and in administration.

R4D will encourage women's participation in road works and emphasise this with contractors and their staff. From experience there it is necessary to work actively on gender encouraging women's participation in construction as the sector is dominated by men. It will be important to understand the barriers to women's participation and to develop appropriate measures to overcome these. Proposed outcome 3 includes an indicative target of 30 per cent female participation in rural road works. This target is based on the experience of TIM-Works and recognises the range of other demands on rural women's time that mean that equal participation is unlikely to be feasible or desirable. The design process will test the appropriateness of this indicative target. The approach to gender issues in R4D will build on the successful practices that have been applied in TIM-Works. To ensure that women have equal opportunity to participate in the planning and implementation of road works, inclusive community engagement and information sharing will be central to the implementation of R4D. Provisions will be included in contract documents outlining requirements for involving women and providing incentives for doing so. Intensive awareness raising campaigns will be organized at various levels among stakeholders. The output-based (*i.e.* task-work) system used under TIM-Works will be further refined and continued as it facilitates the engagement of female workers by allowing flexibility in working hours.

In addition to encouraging women's participation in road works, further consideration and analysis will be undertaken through the design phase (or if necessary following program inception) on the following issues:

- Ensuring women's priorities are taken into account in the planning of road works, including selecting and prioritising roads for rehabilitation and maintenance
- Managing tensions or conflict between men and women arsing from the direct income provided by road works (*e.g.* potential negative reactions to income-earning opportunities for women, insufficient women's control over income, use of income for socially-damaging activities)
- Managing negative gender impacts of road works arising from the presence of outside workers in or near communities (likely to be minimized by the focus on using local workers)
- Promoting women's involvement in the management of local contracting companies.

Child Protection

The design and implementation of R4D will adhere to the ILO policies, including a strict prohibition on child labour, and implement appropriate safeguards for children who may accompany their parents or guardians to road works sites. The ILO's global report on child labour sets out a global action plan to protect children and eliminate child labour in the world of work. This plan gives strategic directions for its operations at the country, regional and global levels and outlines key actions to be taken. The plan provides a clear vision for a world without child labour and member states commit themselves to the elimination of the worst forms of child labour by 2016.

Environmental Safeguards

R4D will be designed in line with ILO policies and AusAID guidance on integrating environment into activity design. The Constitution of Timor-Leste prescribes, amongst other environment objectives, that the "State should promote actions aimed at protecting the environment and safeguarding the sustainable development of the economy". The State Secretariat for the

Environment (SEMA), under the Ministry of Economy and Development, has the responsibility for environment impact assessment. The design of R4D will consider how environmental impact assessment for roads can draw on SEMA's role.

Labour-based approaches are by nature relatively environmentally friendly. However, even with these types of interventions there are environmental impacts. It will be important for R4D to make an assessment of the main impacts caused by rural road rehabilitation works and to provide technical guidance on how to mitigate negative impacts, including an environmental impact assessment methodology that is coordinated with SEMA's approach to impact assessment. Environmental issues will be integrated into R4D capacity building activities. R4D will not only mitigate possible adverse environmental effects related to its rehabilitation works but it will also improve the environmental impact of existing roads by:

- Providing robust designs (including drainage structures and civil engineering and bio-engineering slope-stabilization and protection measures)
- Focussing on improving and maintaining already existing road alignments
- Applying environmentally-sound work methods that make a balanced used of locally available resources and
- Maintaining rehabilitated roads.

Climate Change

Climate change is a significant risk to the long-term sustainability of the road network in Timor-Leste. Australian Government projections of the likely impact of climate change on Timor-Leste indicate that average annual rainfall will increase over the long term. However, these increases are likely to occur in those months with already relatively high rainfall. Extreme rainfall events are likely to become fewer but more intense and therefore may have a more significant impact on the road network.

To enhance the sustainability of the rural road network, the impact of projected higher and more intense rainfall will be considered in the design of both the capacity building and direct investment components of R4D. Consideration will be given to whether modifications to the proposed road works, such as additional drainage or enhanced maintenance, are necessary to manage the expected impact of increased rainfall as a result of climate change. The design will include cost estimates for such modifications. The design will outline the process by which detailed climate risk assessments will be undertaken and integrated into the planning of road rehabilitation and maintenance. As far as practicable, developing appropriate skills and processes in understanding and managing climate risks will form part of the capacity building activities of R4D.

Design Issues and Timelines

R4D will be jointly designed by the ILO and AusAID. The ILO will lead the process and prepare the design document. The design team should include members with the following skills (or combination of skills): labour-based infrastructure works; institutional capacity building; monitoring and evaluation and/or program logic; gender equality and social safeguards; and environment and climate change. GoTL stakeholders will be engaged throughout the design mission and the process will be structured so that they are part of key design decisions. Prior to fielding the design mission it is recommended that preliminary work is undertaken, including bilateral consultations and discussions between AusAID and the ILO in Dili with GoTL stakeholders, and the collection of information that will facilitate an effective design mission.

The following time-line is envisaged for the design process:

- Peer Review and approval of the Concept Note: June 2011
- Design mission and consultations: July 2011
- Finalization Design Document: August 2011
- External Appraisal: September 2011
- Peer Review and approval of the Design Document: September 2011
- Formal Approval by ILO, AusAID and GoTL and mobilization: October-December 2011
- Start implementation activities: February 2012.

This schedule will facilitate a smooth transition from TIM-Works to R4D.

Risks

A number of possible risks have been anticipated in preparing this Concept Note. A risk assessment matrix is outlined below. By building on successfully demonstrated approaches in TIM-Works and other capacity building and rural infrastructure programs, and by coordinating with relevant programs, it expected that the risks can be minimized and mitigated.

An immediate risk to program implementation is the possibility that GoTL's current commitments to R4D may not materialize. This is a key point that needs to be addressed during the design and firm commitments from GoTL need to be solicited. In the longer term, there is a risk that GoTL will not make the increased budget allocations necessary to institutionalize an effective system for rural road rehabilitation and maintenance. They may consider that R4D will provide sufficient funds to this sector and divert their resources elsewhere. This risk will be managed through policy dialogue and by demonstrating the suitability of labour-based methods.

The government's procurement policy has been changing yearly and impacts on the implementation of road works, in particular quality. This risk will be mitigated by building in flexibility within the program to respond to a changing procurement environment and to engage the government in dialogue on procurement policy issues.

Monitoring and Evaluation

A Monitoring and Evaluation (M&E) framework will be developed by the program to ensure accountability and transparency in the use of resources, to monitor progress and achievements, and to ensure that lessons learned in the implementation of R4D are distilled and shared with stakeholders, including the GoTL and donors. The M&E framework will be fully compliant with the reporting, monitoring and evaluation requirements of AusAID. The M&E framework will cover all program outcomes, including GoTL resource allocations, capacity building and direct investments. The framework will measure gender outcomes and adherence to child protection and environmental safeguards.

Key information related to the progress of R4D will be captured by the program's Management Information System (MIS). As much as possible this will use government monitoring systems. R4D will work to strengthen GoTL systems for monitoring the delivery and impact (including gender impacts) of rural road works, including by ensuring that the Roads Management Database developed by MoI (through the support from JICA) captures information about the roads rehabilitated and maintained through R4D.

For monitoring the achievement of the program's effects and impacts a series of studies will be conducted. Base-lines will be established at the beginning of the program and regular follow-up studies will be implemented to enable an evaluation of the performance of the program against its intended results. Detailed impact analysis, disaggregated by gender, will be targeted to selected districts to make efficient use of evaluation resources. Sound M&E systems, including targeted impact analysis, that generate robust information on the benefits of rural road rehabilitation and maintenance and the effectiveness of labour-based methods will contribute to policy dialogue on budget allocations and support to system strengthening.

9

Public Transport

INTRODUCTION

PURPOSE OF THESE GUIDELINES

The Public Transport Guidelines for Land Use and Development aim to assist decision making on statutory and strategic planning proposals for land use developments that affect public transport planning and delivery. It is intended that these Guidelines will assist with site design to facilitate the delivery and use of public transport services. Good design for public transport helps ensure the provision of a sustainable transport network now and for the future. Efficient public transport has a significant influence on the way our cities and regions grow and develop. Land use patterns which are designed to effectively accommodate public transport have a fundamental effect on how well public transport services can be delivered and utilised. The Guidelines will also assist in addressing the public transport aspects of:

- Structure plans and other strategic planning documents
- Integrated Transport Plans (ITP)
- Layout of new subdivisions and major developments.

Why Public Transport Guidelines are Important

The provision and utilisation of a well designed public transport service network will return a range of significant benefits to the community. In many situations, public transport can provide a viable and preferable alternative to travel by car. Providing travel choices for car owners is central to achieving more sustainable transport outcomes. The provision of a well-connected and frequent public transport service also increases opportunities for non-car owners to access employment and other activities. Designing our towns and activity centres to accommodate public transport services, infrastructure and facilities will have additional benefits such as reducing air and noise pollution and road traffic congestion, so as to create more vibrant, walkable town centres. Road-based public transport services should become more reliable and journey times will be reduced. This will be achieved by implementing a range of measures such as allocating priority for trams and buses on congested roads, improved design of bus routes, improved connections and interchange facilities and targeted investments to increase rail capacity. These outcomes are consistent with the Government's social, economic and environmental objectives as outlined in Melbourne 2030, A Fairer Victoria, the Metropolitan Transport Plan and Meeting our Transport Challenges.

Who are these Guidelines for?

This document is designed to assist local government authorities, developers, operators, agencies and consultants in regard to:

- The likely requirements of the Director of Public Transport when performing the Referral Authority role, ie. refer to Clause 52.36 (Integrated Public Transport Planning) of the Victoria Planning Provisions (VPP)
- Clarification of the implications of the State Planning Policy Framework particularly in relation to Clause 12.08 (Better transport links) and Clause 18 (Infrastructure)
- Technical advice on public transport, eg. in preparing permit applications and development plans.

Delivery of Public Transport Services

In most cases the government, in consultation with stakeholders, determines the location and type of public transport services that operate in Victoria. Route-based public transport within Victoria is delivered by the private sector under contractual arrangements with the state government.

These Guidelines aim to ensure that new developments are designed and implemented in ways that support public transport operations. However, the decision on the level of public transport service provided at a particular site is made as part of the budget processes of government.

PRINCIPLES FOR PROTECTING PUBLIC TRANSPORT

In the past, some aspects of developments have significantly impacted on public transport operations. In responding to proposals, either as a referral authority or as a party who may suffer material detriment, the Director will act to protect public transport operations and infrastructure. Proponents must therefore design proposals so that they adhere to the following principles:

- New level crossings should be avoided. Urban connectivity should be maintained or achieved by grade separations for pedestrians and vehicles. The design of transport routes at new developments must provide for grade separation at level crossings except with the approval of the Minster for Public Transport.
- The proposal must not delay public transport services (except through increased patronage) or otherwise detrimentally impact on public transport operations. New signals should be avoided where

possible, as these usually result in delays to public transport, despite signal phasing priority for buses and trams. Proposals which introduce even small detrimental impacts can cumulatively impose significant delays on public transport services.

- Proposals should identify potential impacts and details of how those impacts will be mitigated. Private proposals should not require public investment to redress delays they cause.
- Proponents should try to avoid locating service facilities across public transport corridors, as drains, etc, add to the cost of developing public transport infrastructure. Where service crossings will be unavoidable, VicTrack and/ or the Public Transport Division of DOT (PTD) should be contacted. Proponents are encouraged to consult with PTD at an early stage.

When to Contact Public Transport Division (PTD)

Public transport planning should be an integrated part of site planning. Addressing pedestrian and mobility access considerations at the time of initial site planning is crucial to integrated transport and land use planning. This will ensure that public transport is not an add-on but is considered as part of the design process.

If public transport is not considered until the end of the design process, this can result in sub-standard solutions for public transport operators, poor outcomes for transport users and reduced access to a development. PTD is responsible for the public transport system and can advise on network plans and design principles. When contacted, PTD will provide up-to-date information of the Government's system-wide and site-specific public transport developments, as well as informal feedback on concepts as they are developed. Before submitting a planning permit application to the council as responsible authority, please contact the Land Use and Planning Referrals Team, PTD Department of Transport.

Referral Process

The Director of Public Transport is a Referral Authority pursuant to Section 55 of the Planning and Environment Act 1987. Details of requirements for referrals can be found in Clause 52.36 (Integrated Public Transport Planning) of the VPP. This means that development proposals which are above the thresholds set out in Clause 52.36 must be referred to the Director for his comment. A Referral Authority must consider every application being referred and may tell the responsible authority in writing that:

- It does not object to the granting of the permit
- It does not object if the permit is subject to conditions specified by the referral authority
- It objects to the granting of the permit on specified grounds.

If the Director of Public Transport receives an application and considers that insufficient information has been provided to undertake an assessment of the proposal, then further information may be requested of the applicant. If the applicant does not agree to the request for further information, an Application for Review may be lodged at the Victoria Civil and Administrative Tribunal.

The principles, design goals and requirements set out in the Guidelines are used to assess planning permit applications referred to the Director of Public Transport. Clause 52.36 also sets out circumstances where a referral will not be required, eg. a referral of an application to the Director of Public Transport will not be required where a development is consistent with an adopted structure plan prepared in consultation with, and endorsed by PTD. A variety of public transport requirements need to be taken into account when undertaking major developments and these are assessed on a case-by¬case basis.

Prior to preparing and submitting planning permit applications, developers are encouraged to liaise directly with PTD for major development proposals.

PLANNING AND DESIGNING FOR PUBLIC TRANSPORT

PUBLIC TRANSPORT NETWORKS

The public transport network in metropolitan Melbourne is made up of two elements:

- The Principal Public Transport Network (PPTN) which is the passenger rail network, the tram network and selected metropolitan bus routes.
- Local networks and services, such as local bus routes, taxis and school bus services which connect to and supplement the PPTN.

Rural and regional public transport services comprise a network of core rail and coach services connecting regional centres and metropolitan Melbourne. Local services include local bus, taxi and school bus services. Public transport modal interchanges provide for transfers between motorised and non-motorised transport modes, including cycling and walking. These are generally located in Activity Centres.

PRINCIPAL PUBLIC TRANSPORT NETWORK

The PPTN will be the focus for the development of a high quality public transport system that connects the Principal, Major and Specialised Activity Centres. The on-road component of the PPTN includes the tram network and a number of major cross-town and radial bus routes. The off-road component of the PPTN includes all metropolitan passenger rail lines. The PPTN includes both existing routes and anticipated future routes. By locating major developments on the PPTN, opportunities for integrated development and enhanced public transport outcomes will be maximised.

Improvements sought for the PPTN aim to achieve a reduction in travel times and an improvement in service reliability for public transport services. Integrated land use and transport planning for the PPTN should include:

- Land use decisions to enhance access to the PPTN
- Managing road space giving priority to public transport in congested areas to improve the speed and reliability of tram and bus services
- Facilitating public transport services through interchanges
- Improving the convenience and safety for pedestrians accessing the PPTN
- Upgrading amenity at nodes and transfer points of the PPTN. The list of existing, proposed and possible PPTN routes may vary from time to time. Local government should also provide input to the planning of future transport routes, particularly those that support the PPTN. Contact PTD to obtain information on current and potential PPTN routes. These routes should be considered when planning developments.

Local Network Buses

While the PPTN plays a key role in longer-distance travel, in many cases, access relies on the local public transport network. Most local routes are provided by buses.

Local network bus services access local neighbourhoods, Neighbourhood Activity Centres, Activity Centres and Railway Stations, where interchange to services on the PPTN can take place. The expectation for Neighbourhood Activity Centres is that they will all be connected to the regional Major Activity Centres (MAC) and/or Principal Activity Centres (PAC) by local public transport services. However, in some cases this role may be also performed by PPTN services.

The ability to provide both direct local services and comprehensive service coverage for a local area will also be based on functionality of urban form, road layout and the level of recurrent budget support. Improvements to local network services are being made, in addition to premium cross-town services and additional routes and local feeder services. Specific improvements include demand-responsive services, the SmartBus program and vehicle design enhancements.

Regional and Rural Services

The major regional centres close to metropolitan Melbourne (Geelong, Ballarat, Bendigo and the towns of the Latrobe Valley) are experiencing accelerated development. Initiatives for these "networked cites" include investigation of transport corridors and service improvements.

Fig. Box Hill Transport Interchange

NETWORK PRINCIPLES: PUBLIC TRANSPORT AND LAND USE

Targets for planning and designing new public transport routes include the following:

- Ninety-five per cent (95 per cent) of residential land uses in established and urban growth areas to be designed to allow access to public transport services within 400 metres safe walking distance.
- To improve accessibility for the disabled and elderly, it would be desirable that aged-care facilities, educational, medical and community facilities be located within 200 metres of an existing or planned bus stop, train or light rail station or tram stop. Site factors such as steep grades may reduce this distance. In cases where public transport facilities need to be accessed by patrons with limited mobility, appropriate facilities should be located on the route
- As a guideline, bus and tram stops should be located every 300 metres and reflect the location of key attractors
- Every home within urban areas should have direct access to a Principal or Major Activity Centre by public transport, ideally with a maximum travel time of 30 minutes without changing vehicles
- All Principal and Major Activity Centres should be located on the PPTN.

ACTIVITY CENTRE PLANNING AND PUBLIC TRANSPORT

Activity centres operate as key nodes in the public transport system. The Activity Centre Design Guidelines (DSE) note that:

"If activity centres are to fulfil the promise of increased sustainability, they need to attract more public transport users. Many activity centres function as critical nodes in the public transport system already. However, if public transport use is to be increased, it must become a central feature of every activity centre and offer a more convenient alternative to car-based travel. Improving the connection between different routes and modes of public transport in well-located interchanges that are integrated with the street and building fabric of the activity centre is an important component in encouraging increased use."

Activity Centres

Planning for activity centres should improve movements of public transport vehicles, provide for convenient, safe and user friendly interchanges and ensure an appropriate level of integration with the surrounding streetscape, street network and built form. Objectives for integrated planning of public transport and activity centres include:

- Encouraging public transport use by providing convenient, prominent and active environs around railway stations and interchanges.
- Providing high quality amenity for passengers (including lighting, seating and shelter)
- Providing safe, attractive and direct pedestrian and cycle access to railway stations, interchanges and transit stops
- Minimising the potential dividing effect and urban form impacts of railway corridors
- Supporting the reduction in the need for car parking where appropriate.

Fig. Subiaco Railway Station - WA

Activity Centre Structure Plans

The involvement of PTD in the creation of structure plans is a crucial part of activity centre planning. PTD is keen to support local government to identify strategic directions and proposals for activity centres, which will support the use, operation and delivery of public transport services. Structure plans prepared for activity centres should reflect the following principles:

- Incorporate higher density mixed use development near public transport nodes which in turn will support increased public transport services, extended hours of operation and higher frequency levels
- Surround railway stations, transit stops and interchanges with active, ground-level uses
- Integrate transit stops and interchanges into the design and layout of the activity centre, and locate them centrally
- Prioritise pedestrian access in the activity centre and to the public transport modal interchange whilst also facilitating easy movement of public transport vehicles
- Recognise that the public transport interchange is a 'gateway' to the activity centre
- Design for minimal access by private motor vehicles into the modal interchange, to reduce competition for road space, remove conflict between buses, cars, pedestrians and cyclists and maintain reliability of public transport services
- Provide direct access for public transport vehicles to minimise travel times
- Give priority to public transport movements, especially on the PPTN
- Provide "kiss and ride" and taxi facilities around transport interchanges. For more information regarding structure plans and activity centres, please refer to Activity Centre Design Guidelines (DSE). Reference should also be made to the Planning Practice Note – Structure Planning in Activity Centres.

Growth Areas Precinct Structure Plans

The DSE Precinct Structure Planning Guidelines for creating new communities in Melbourne's Growth Areas set out the objectives for designing new neighbourhoods. The planning principles must be met: Objective 3: To provide sustainable transport networks: "Early planning of public transport services, including local bus services is critical and should be a key element that defines the structure of new neighbourhoods".

PTD Structure Planning Process

Early involvement of PTD in the structure planning process is highly desirable. PTD can assist the preparation of structure plans in a number of capacities including:

- Providing advice and feedback on proposals to increase access to and use of public transport services
- Providing advice on improvements to public transport services and facilities including longer term network improvements and operational requirements

- Providing advice on Government's priorities for the future planning and delivery of public transport services and facilities
- Assisting in consultations with public transport service providers. A coordinated approach to the implementation of structure plans is critical in integrating land use planning with transport planning. Structure plans should clearly identify responsibility for delivery, proposed timing and sequencing of actions.

Referral of a permit application (under Clause 52.36 of the VPPs) is not required if the development is consistent with an adopted structure plan that has been prepared in consultation with and endorsed by the PTD.

SUBDIVISION DEVELOPMENT AND PUBLIC TRANSPORT

Subdivisions must identify and make provision for public transport routes and pedestrian and cycle access to public transport services. Public transport will need to be a core consideration in subdivision design. Planning to provide access by public transport should be integrated with the preparation of development plans and planning permit applications. These Guidelines provide technical design guidance for the provision of public transport access and infrastructure. New urban developments should be contiguous, allowing sequential public transport route development.

RELATIONSHIP BETWEEN TRANSPORT ROUTES AND THE SURROUNDING AREA

Pedestrians

Public transport networks and services should be considered in the context of their surrounding areas, including how pedestrians will access the services. This means that pedestrian design objectives such as Disability Discrimination Act (DDA) accessibility, safety, amenity and urban design considerations are important when planning for public transport. High standard facilities for walking will support an increase in public transport patronage. Paths should be legible, smooth, uncluttered, well lit, overlooked by active land uses, have places to rest and include safe crossing points. Key design principles include:

- Paths should be separated from traffic, direct and of sufficient width (particularly if the path is shared by pedestrians and cyclists)
- Provide landscaping and street furniture for shade, comfort and street enclosure
- Pedestrian crossings should be conspicuous and clearly indicate where pedestrians can cross *e.g.* include zebra crossings, use of colour or raised pavement to delineate pedestrian priority space
- Main pedestrian routes and public transport stops should be well lit, have good visibility and be DDA compliant

- Car parking spaces for the mobility impaired should be very close to interchange entrances
- Maps and way-finding signage should set out key destinations, including public transport stops
- Public transport, walking and cycling routes should connect to the heart of an area and be as close as possible to meeting areas and public spaces
- Pedestrian and cyclist access to public transport should be convenient and direct. Similarly, access between bus, rail and tram should be direct.

Pedestrian Crossings

In new railway and light rail stations, pedestrians should have separate, safe routes to the platform. Paths should be planned along pedestrian desire lines and not require pedestrians to walk along car park aisles or driveways. The following design principles apply to the design and designation of a pedestrian crossing within or to an interchange facility:

- Crossings which require pedestrians to turn their heads at a greater than 45° angle to view oncoming buses and trams are unacceptable.
- In many situations, a 'zebra' type crossing provides sufficient control and protection for pedestrians
- Signal controlled crossings are necessary where high pedestrian flows exist
- Standard crossing width is 2.4 metres
- Crossing length should not exceed 10 metres, unless refuge islands are

Cyclists

Cyclists need to ride safely and easily to and from public transport facilities and along roads used by buses. Bicycle facilities should be considered in all proposals, including changes to streetscapes. Key design principles include:

- Providing a mix of safe bicycle parking and encouraging provision of end of trip facilities in developments and public transport facilities such as railway stations and modal interchanges.
- Signposting key destinations on off-road bicycle paths including distance markers, to provide reassurance and certainty to users
- Installing bicycle head start boxes at traffic signals
- Providing dedicated bicycle lanes/paths to stations and transit stops
- Ensuring all bike networks connect to activity centres, public transport stops and interchanges.

DOT is providing bicycle lockers and cages at railway stations across the metropolitan area. New or refurbished interchanges must ensure adequate space is designed into the layout of the interchange to accommodate bicycle lockers and cages.

Fig. Individual lockers – Watergardens Railway Station – Melbourne

Fig. Bicycle head start boxes

DOT Walking and Cycling Branch

A Walking and Cycling Branch has been established in DOT, recognising not only the importance of walking and cycling as travel modes but also of the potential to grow substantially as the community seeks more sustainable forms of transport. The Waking and Cycling Branch overseas the successful TravelSmart travel behaviour change program and the Local Area Access Program, which supports local government in developing a range of demonstration projects that will improve local access through walking and cycling. The Branch works closely with a range of stakeholders, including PTD, VicRoads, other state government departments and agencies, councils, cycling and walking organisations (including Bicycle Victoria, Kinect Australia and the Cycling Promotion Fund.) This ensures that walking, cycling and local access issues are strategically managed and that appropriate policy and programs are developed and delivered for the Principal Bicycle Network (PBN) and the Municipal Bicycle Network (MBN).

"Park and Ride" Facilities

"Park and Ride" facilities enable passengers to use private vehicles to access public transport facilities and therefore divert long car trips onto public transport. They also enable an expansion of the catchment for the public transport network. Strong growth in public transport usage is reflected in the increasing demand for parking spaces at railway stations. There is pressure to increase the size of "Park and Ride" facilities. Meeting Our Transport Challenges (MOTC) recognises this demand and includes a substantial program to increase the number of park and ride spaces. The program is being developed and delivered by PTD.

Planning for "Park and Ride" facilities should reflect the following:

- Careful design is required and should include generously proportioned accessways and landscaping
- Very large park and ride facilities may be more appropriately located at railway stations on the periphery of the key activity centres, rather than at the activity centres themselves.
- Private vehiclesshould not obstruct access to the station interchange for buses, pedestrians and cyclists
- Private vehiclesshould not, but may attempt to use bus interchanges for passenger drop-off/ pick-up
- Commuter car parking facilities should be near to station entrances and should not compromise pedestrian or bicycle access.
- Security and safety are prime concerns at commuter car parks
- A "Park and Ride" should include a "kiss and ride" area at all key activity centres
- The size of each "Park and Ride" facility will depend on local demand
- Excessive car parking at railway stations in the key activity centres may compromise the successful functioning of centres, lessen the focus on the station as a key element of the activity centre and be unattractive.

DISABILITY DISCRIMINATION ACT (DDA) COMPLIANCE

Works undertaken on the public transport network are required to comply with the Commonwealth Disability Discrimination Act 1992, and particularly the technical requirements of the Disability Standards for Accessible Public Transport 2002 (DSAPT). Developers should carefully check their responsibilities and liabilities under the Act at the earliest stage when contemplating a project. If a private development will affect public infrastructure, it may trigger a requirement for the developer to make the infrastructure fully compliant with the Act and the DSAPT.

All new public transport services and facilities coming into operation after 2002 had to comply in full with the DSAPT. All existing works and services had to be retrofitted or replaced to comply with the milestones of 25 per cent compliance by 2007, 55 per cent by 2012, 90 per cent by 2017 and full compliance by 2022. Trains and trams (vehicles) must comply in full by 2032. The DSAPT applies to all conveyances, premises and infrastructure. There are requirements for circulation, ramp grading, handrails, tactile ground surface indicators, toilets, furniture, information, signs, lighting, etc. Vehicles include trains, trams, route buses, scheduled coaches, taxis, ferries and aircraft. Premises and infrastructure include train stations, pedestrian rail level crossings, tram and bus stops and taxi ranks, jetties and other customer service points. School bus services are exempt from the requirements of the DSAPT and local arrangements are often made for people with disabilities under the DDA.

When implementing works, accessibility for passengers with all types of disabilities is a necessity. This includes the requirement for all ramps other than kerb or step ramps to be of grade 1:14 or less and for shelter and boarding points to be equally accessible to patrons of all abilities. For kerb or step ramps the maximum grade is 1:8.

Fig. Out-of-Centre Park and Ride – Doncaster

Fig. Park and Ride - Werribee Railway Station

Upgrading tram stops is complex, involving many stakeholders including VicRoads, tram service providers, local government, property owners and the general community. Platform stops are the preferred DDA compliance solution providing faster and safer loading for all passengers.

Achieving seamless access to train, tram and bus services may require integration of local government and private sector infrastructure.

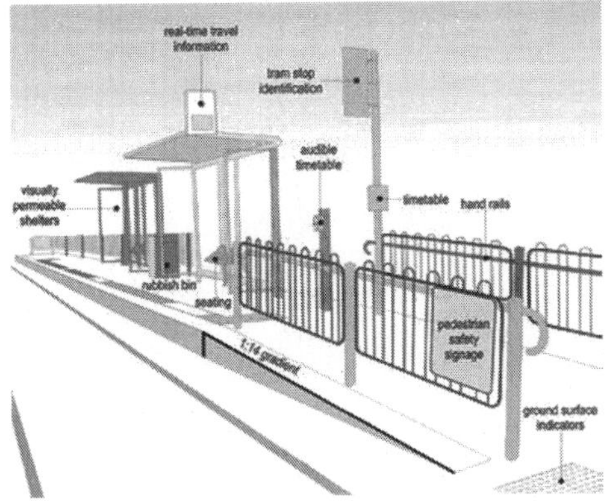

Fig.DDA compliant tram stop

BUSES

Bus operations and requirements must be carefully considered. Local bus networks seem relatively flexible, however in practice, it is often difficult to alter a bus route once introduced, given community expectations, established travel patterns and the cost of stops and signs. Therefore, it is important to ensure that the bus route and network planning are correct from the outset. It is important to ensure that routes are connected and users can transfer easily between services to access a wide range of activities. Developers should liaise with PTD to discuss current and potential future bus networks, operational arrangements and how they integrate with proposed developments. Consultation with bus operators can also provide local information on routes and operations. New developments will need to reflect the following, so as to accommodate new or revised bus routes and their supporting infrastructure:

- Road Rules – Victoria
- PTD advice and requirements regarding existing, proposed and potential bus routes, as well as bus infrastructure, stops, priority and bus layover areas
- VicRoads' Bus Stop Guidelines and Bus Priority Guidelines, VicRoads' Traffic Engineering Manual, Volume 1 and VicRoads' Designing Local Roads for Ultra Low Floor Buses Bulletin.
- Design that minimises delays, so that bus services can achieve contracted operational performance targets

SERVICE PERFORMANCE REQUIREMENTS

Bus Catchments

Land use developments should propose public transport routes in locations which will optimise population catchments and service viability. Some principles for efficient operation and maximising the demand for services include:

- Neighbourhoods should be designed for bus routes on strategically located connector roads so that dwellings will be within 400 metres of a bus route. For example, if arterial roads are spaced 1600 metres apart, a bus route along a 'central' strategic connector would place virtually all persons within 400 metres of a bus route (either on the connector or arterial roads). Layout of optimal bus catchments is illustrated in Figure.
- Design should allow for direct routes and easy access between key destinations
- Higher density development should be located along public transport routes
- Bus stops should be provided in convenient locations
- Direct and safe pedestrian access should be provided to bus stops
- Provide public transport and pedestrian access at and within key destinations such as modal interchanges and activity centres that is suitable for high volume and high frequency passenger movements.

Fig. Bondi Junction - Bus and Train Interchange

New Subdivisions

Bus route planning for new and/or extended bus services will require sub-regional or development area planning. PTD will assist local government and developers to incorporate wider bus networks and service strategies into their proposals.

Specific elements to consider when planning new land use development for all uses include:

- Patronage-generating uses should be located on both sides of the bus route, ie, avoid 'dead areas' within or between subdivisions where no demand generating land uses abut the route
- Provide direct road connections designed to accommodate buses, as outlined in these Guidelines and in Clause 56 of the VPP (Sustainable Neighbourhoods).
- Developments incorporating community facilities (*e.g.* schools, active recreation grounds and tourist facilities) that will or may require bus and coach access should be located to take advantage of existing public transport services or so that aggregate demand for public transport is optimised.
- New infrastructure investment should aim to enhance existing operations and safety
- Stage development in sequence, to allow efficient deployment of bus services.
- Stage the extension of public transport services and provide interim service arrangements until the services can be extended
- Consider the effect of the new development on services moving through or past the development
- The provision of through routes helps to ensure the efficiency and attractiveness of bus services
- Stop infrastructure should only be installed immediately prior to route commencement and should be fully DDA compliant.
- Where it is deemed that a bus route should not pass through a subdivision, the developer should ensure appropriate pedestrian linkages are provided to access existing routes.

DESIGN PRINCIPLES (BUSES)

Local Road Design

In new developments, the following road design specifications are required for bus operations. For further information refer to VicRoads' Designing Local Roads for Ultra Low Floor Buses Bulletin.

Table. Road Design Specifications for Bus Operations

Characteristic	Specification
Limit of acceptable grade %	
Maximum grade (in limited circumstances where acceptable alternatives are not possible)	9%
Minimum distance between legs of staggered 'T' intersection	40m

Pavement design	Heavy vehicle frequency, including buses, will dictate required pavement strength.
Pavement appearance – traffic and or parking	Areas for traffic and indented parking need to be clearly distinguished. Areas intended for parking should have marked bays.

The trafficked road width of any carriageway, including one way roads, needs to be sufficient for two buses to pass at normal speed. This width excludes any bicycle lanes, parking lanes/spaces and manoeuvring space for parking vehicles and any traffic islands. See below for more details.

On road sections, other than through roundabouts, undivided connector roads to be used by buses must have:

- For two lane, two way roads a clear trafficable road width of 7.0 metres, with separate designated space for cyclists and/or parking. If parking is intended, then a separate parking lane or indented spaces, or a shared parking and bicycle lane should be provided. Note that the cycling and parking spaces are in addition to the 7.0 metre trafficable road width required for bus operations, as shown by the Undivided Connector Road - A in Figure.
- Alternatively, in situations where the above is not feasible, traffic lanes for shared motor traffic and bicycle use of 4.2 metres (minimum) should be provided. If this arrangement is provided on both sides, a trafficked road width of 8.4 metres is required, plus any space required for parking, as shown by the Undivided Connector Road - B in Figure. Any parking must be provided in a parking lane of 2.3 metres (minimum).

Connector streets with a central median, if to be used by buses, must have a road carriageway width of:

- At least 5.0 metres in each direction, when a shared bicycle and traffic lane is provided
- 5.2 metres when a dedicated bicycle lane and a separate traffic lane are provided.

In conjunction with the above carriageway dimensions, the overall clear zone must be 7.0 metres. The median surface within the clear zone must be trafficable by low floor buses in all weather conditions and the kerbing must be mountable or semi-mountable in accordance with road authority guidelines.

The specified width of the clear (and trafficable) zone is to ensure that a clear path is available for through buses if a vehicle stops, stands or breaks down within the one way road section. Boulevard treatments are typically urban design features, used to create an entry to a development and are usually short in length. No lanes, other than a bicycle lane, should be included within the 7.0 metres width. Parking facilities must be in addition to the 7.0 metres width. In all cases, road design and any traffic management features must be able to accommodate the turning templates for the standard 12.5 metre ultra low floor bus. The 12.5 metre ultra low floor bus is the standard for DDA compliant bus routes in Victoria.

Design for 19 metre articulated buses is required on existing, proposed or possible articulated bus routes, including access routes to activity centres which may in the future be served by articulated buses (PTD and operators should be consulted).

For roundabouts, motor traffic lanes into, through and for departure, need to be at least 4.0 metres wide. The taper to this increased lane width must achieve the 4.0 metres minimum lane width by the tangent point where the kerbing curves for the outside of the roundabout and be otherwise designed in accordance with Austroads Guidelines. Where bus movements will involve turns, to ensure pedestrian safety, sufficient road pavement must be provided within intersections to accommodate the swept path of the bus without the front or the rear overhang of a bus sweeping over any pre-existing splitter island or kerb – particularly if the rear overhang sweeps over a surface that could be used by pedestrians.

On roads that may occasionally be used by buses, to accommodate infrequent left-turns by ultra low floor buses, the surface of splitter islands is to be less than 0.1 metre above the road surface.

Splitter islands should also be clear of obstructions, except that a 'roundabout' sign may be located off centre in the side of the island adjacent to approaching traffic and positioned 0.5 metre behind the back of kerb of the outer edge of the roundabout.

Local Area Traffic Management

Connector roads with 60 km/h or less speed zones are the preferred local road class for buses and are not usually built with or retrofitted with Local Area Traffic Management (LATM) devices. Where LATM devices are being considered on existing and future bus routes, impacts on bus travel times, the safety and comfort of passengers, and driver fatigue should be considered. Drivers need to negotiate the devices, with little room for error, many times each day along the length of each route.

Some common major and minor traffic control items have a serious impact on bus operations and should be avoided, including:

- Speed humps or raised platforms
- Mountable or semi-mountable roundabouts
- One-way road narrowing/slow points and 'weave points'. In new developments, the street network should be designed with a focus on traffic calming to achieve desired speed limitations, rather than relying on LATM devices

Road Humps and Speed Cushions

Designers should have regard to the following matters.

- No new road humps (and raised platforms) should be constructed on roads which form part of a regular bus route
- Speed cushions, which are intended to allow large vehicles (including buses) to straddle them, while requiring smaller vehicles to slow down, should only be installed on smaller access ways (*e.g.* loop roads around a neighbourhood centre). The use of speed cushions also requires consideration of cycling and parking arrangements.
- Speed cushion installations on bus routes must be designed to ensure that buses can straddle the cushions without experiencing any bump and without requiring a bus to travel too close to the kerb or road centreline. Generally, the outer cushion base must start no closer than 0.8 metre to the face of the kerb and not require the bus to deviate significantly from its path to straddle the cushion. Note that on roads with marked bike lanes, cushion placement will need to be designed to allow for cyclists, slow cars down and provide a smooth passage for buses. Speed cushions must be provided in accordance with VicRoads' Traffic Engineering Manual Volume 1 and Austroads' Guide to Traffic Management, Part 8 – Local Area Traffic Management
- Street lighting must be provided at speed cushions

Roundabouts

Roundabouts are not favoured on bus routes as they introduce delays (ie. buses are subject to the same "give way" rules as other vehicles).

Slow Points

Two-lane parallel and angled slow points are the only appropriate forms of slow points for roads used by buses. A minimum lane width of 7.0 metres should be provided (excluding bicycle and parking lanes) at a parallel slow point. Greater width is required for angled slow points, with each case being different. Examples are shown in Austroads' Guide to Traffic Management, Part 8 – Local Area Traffic Management and VicRoads' Traffic Engineering Manual Volume 1.

One-lane, two-way slow points (parallel or angled) are not to be used on bus routes because of safety concerns and delays to services. Offset and double offset slow points, based on the "one lane" principle, are also inappropriate. Parking restrictions may be required near slow points to accommodate the swept path of buses. Slow points should be coordinated with bus stops to minimise delays and adequate street lighting should be provided for night time visibility.

Fig. One-lane, two-way slow points are not acceptable

Splitter Islands (at Locations other than at Intersections and roundabouts)

Lane widths and islands should be in accordance with Austroads' Road-Based Public Transport and High Occupancy Vehicles. Landscaping and beautification of the islands should not reduce sight distances.It is not acceptable to require buses to cross a mountable splitter island (because of the discomfort for passengers and drivers). At sites where buses can be expected to turn (*e.g.* at significant attractors or destinations) the swept path of buses must be considered and a path provided on the road surface, clear of any splitter island or kerbing. At other sites where buses may occasionally turn, it may be acceptable to provide for the front or rear overhang of the bus to sweep over kerbing and splitter islands where these surfaces are designed to be less than 0.1 meter above the road surface (to accommodate ultra low floor buses). Adequate street lighting should be provided for night time visibility.

'T' Intersection Deviation

Bus routes should be designed to ensure buses do not make right-turns into major roads unless at a signalised intersection. Design shall be in accordance with VicRoads' Traffic Engineering Manual Volume 1, Chapter 8. Lane widths shall be in accordance with Austroads' Road-Based Public Transport and High Occupancy Vehicles. Bus turning templates should be used to assess the design. The overhang must be clear of the kerb. The template is required to assess all turns and bus vehicle movements. Adequate street lighting should be provided for night time visibility.

On-Road Bus Stop Design (and DDA Compliance)

Bus stops must comply with the Commonwealth Disability Discrimination Act 1992 and the Disability Standards for Accessible Public Transport (DSAPT) 2002. The design of all bus stops should be in accordance with VicRoads' Bus Stop Guidelines and the DOI Requirements for Bus Stop Compliance.

Fig. On road stop design - Wattle Park Primary School

Bus Stop Identification

Bus stops should be clearly identified and provide timetable and route information in accord with PTD and Metlink requirements. The bus stop post and flag must be clear of the clutter of other street furniture, and point away from the road. The post and flag should be in front of other obstacles, and be reflective.

Fig. Bus stop identification

Tactile Ground Surface Indicators

The layout and specification of Tactile Ground Surface Indicators (TGSI) must be in accordance with the DDA DSAPT 2002 Part 18 – Tactile Ground Surface Indicators and should meet Australian Standard AS1428 Design for Access and Mobility Part 4 (1992), Tactile Indicators. Typical layouts are shown in VicRoads' Bus Stop Guidelines.

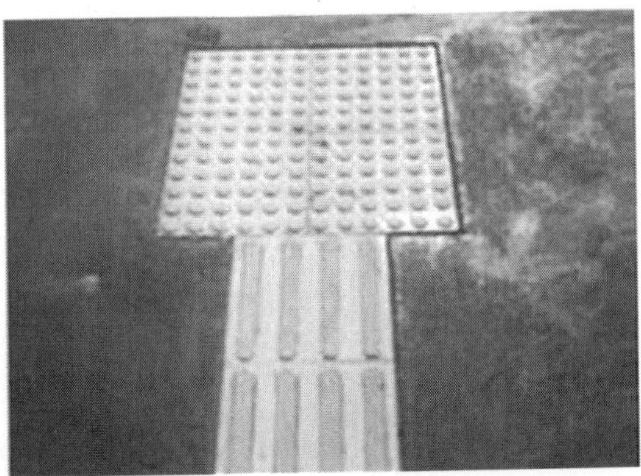

Fig. Tactile ground surface indicator

Bus Passenger Shelter

Passenger shelters should be provided at all stops that are expected to be frequently used (eg. at activity centres, interchanges, community facilities, etc). Bus shelters should be positioned so that the sight distance requirements for road users are not significantly impacted and orientated to provide weather protection. This is illustrated in Figure. Stops without shelters should have sufficient space for a shelter to be provided, at a future time, in accordance with these Guidelines. Bus shelters should be located outside the roadside "Clear Zone" (as defined in VicRoads' Road Design Guidelines). However, if this is not possible, shelters should be constructed of materials that minimise potential injury if hit by an errant vehicle.

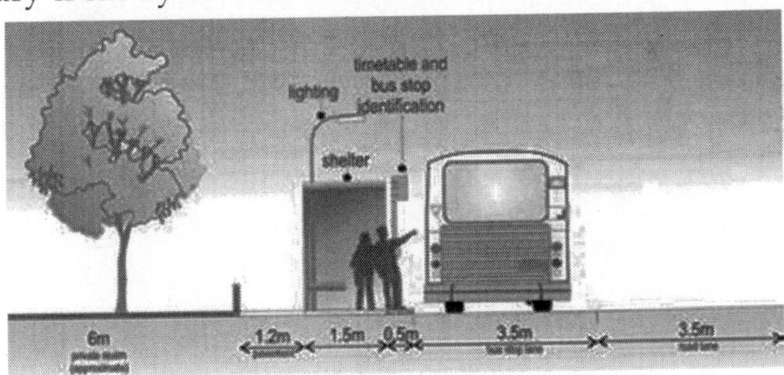

Fig. Bus passenger shelter requirements

Installation of a bus shelter at an existing bus stop will trigger the obligation to upgrade the existing bus stop to full DDA compliance, including all circulation, grading, paving, TGSI and other requirements.

Other Design Considerations

Bus shelters should be located such that passengers are visible to the bus driver and buses are visible to waiting passengers. Lighting should be provided (if streetlights are inadequate) and positioned so as to illuminate waiting passengers in poor light. (eg. the morning peak period in winter commences before sunrise). Shelter size is dependent on the travel demand at the location. Shelters should not obstruct the view of motorists leaving driveways and side streets. When planning new subdivisions, the provision of shelters for bus-stops should be a priority over the location of a new driveway. A 1.2 metre continuous accessible path should be provided behind the bus shelter so as to meet DDA requirements. When this is not achievable, an open ended bus shelter without seats should be provided. If an off-road shared path is provided along the corridor, this path should be maintained at a minimum of 2.5 metres behind the bus shelter. Shelters should not be located in a way which would compromise DDA compliant circulation.

Other Street Furniture

Other street furniture should be located such that the boarding and alighting clear areas are maintained and the 1.2 metres continuous accessible path connects to the pedestrian network. All street furniture should be set back from the kerb by 0.5 metres to allow for bus overhangs. Clear space is also required for both rear and front doors. Bus stops near cycle routes should consider bicycle parking.

Lighting

Lighting at bus stops is to meet the minimum lighting standard requirements of the Public Lighting Code A/NZS 1158. Lighting levels above those specified in the Code should be considered at locations where there is a high demand for the service. Lighting at bus stops should be designed so that the timetable is well illuminated and can be read at night.

Fig. Weather protection from prevailing weather

On-Road Bus Stop Siting

Bus stop siting is critical to maintaining pedestrian and passenger safety and convenience, as well as ensuring ease of bus vehicle access and operation.

The following should be considered:

- Stops should be located as close as possible to demand generating land uses
- The desirable average stop spacing is 300 metres
- Bus stops should be located to maximise walking accessibility from the surrounding area (eg. Near intersections), with particular reference to demand generating land uses
- Where two bus routes intersect, bus stops should be located close together
- On roads with indented parking, bus stops should be designed and located at kerb outstand areas (refer to VicRoads' Cycle Note 20 – Providing for Cyclists at Bus Stops)
- Existing and proposed bus stop locations must be clearly shown in development layout plans
- Bus stops should afford a safe road crossing opportunity reflecting the volume of passengers at the stop and road traffic conditions throughout the day and night. The best case is a controlled pedestrian crossing.

Fig. Continuous accessible pedestrian

- Bus stops used by late night services, such as Nightrider, should include additional safety measures such as bright lighting, overlooking development and a safe road environment that considers the behaviour of substance impaired pedestrians. These may need to be located near taxi ranks and developments expected to generate patronage later at night (*i.e.* hotels, bars and nightclubs)
- Guidance on the selection of pedestrian crossing treatments at bus stops can be found in the VicRoads' Bus Stop Guidelines and the VicRoads' Traffic Engineering Manual Volume 1
- In areas with large trees, structures, etc. close to the road, the vertical clearance needs to be checked, based on the swept path of the bus, to ensure buses are not impeded

Fig. Bus stops located near demand generating uses - High Street - Preston

- Departure side stops at local street intersections are preferred
- Stops should be located away from a major exit point from a development. If safe, design the exit to reduce the speed of entering traffic to reduce the risk of rear end collisions with buses. Buses, shelters, flags/ totems and waiting passengers at stops that are upstream of major access points can obstruct the safe sight distance for drivers trying to enter the road
- Mid-block locations for bus stops should be avoided unless
- They service a particular site
- Passengers can access a signalised pedestrian crossing on a major road, or
- Where blocks are long and mid block stops are required to provide a reasonable stop spacing. Kerbs should be extended past parking bays at such locations, with the bus stopping in the through lane
- Locate stops near existing subways, bridges, school crossings and other pedestrian crossings, to provide safe road crossing opportunities
- Stops should not be proposed near the crest of a hill, due to poor sight distances and the possibility of rear end or vehicle-pedestrian accidents
- Stops on slopes or at the bottom of hills should be avoided because of the difficulty and hazard of breaking and/or accelerating amid general traffic
- Urban/rural fringe stops must also be in accordance with VicRoads' Bus Stop Guidelines.
- Stop siting should reflect the wider environment, including vehicle speeds and safety for slowing, stopped and accelerating buses, pedestrians walking to and from the stop, and other amenity and roadside issues and hazards (*e.g.* shoulder widths, deep drains, long grass, personal safety, etc.)
- Locate bus stops on opposite sides of the road. A site-by-site analysis will need to consider pedestrian movement desire lines

- Bus stops should be downstream from formal mid-block crossings. If several routes share the stop, the location of the stop should reflect the effect of two buses arriving and the pedestrian crossing being temporarily blocked by a bus that is waiting to use the stop
- The stopping restrictions for mid-block islands, which help pedestrians cross busy roads, are regulated in Road Rules – Victoria, and must be considered in conjunction with the stopping restrictions that apply to bus stops
- Pedestrian routes to and from stops should be constructed so that they are safe underfoot, defined and navigable in poor weather and/ or visibility, well lit and landscaped, subject to surveillance and have connections to the broader local movement network and nearby trip generators
- All stop design and location details must be in accordance with VicRoads' Bus Stop Guidelines.

Providing for Right-Turning Buses

A bus at a kerbside stop that must turn right at an intersection, may have difficulty reaching the right hand lane of a multi-lane road. The following solutions may be applied:

- Locating the bus stop downstream of the intersection on the road the bus intends to turn right into
- If road geometry allows, create a separate right-turn lane with associated bus stop facilities
- Depending on traffic flows, a stop located further upstream may provide bus drivers with additional distance, time and opportunity to merge into the right lane.

Future Bus Stops

Proposed bus stops must provide hard-stand areas as specified in these Guidelines, with sufficient space for the future installation of the appropriate requirements identified in the VicRoads' Bus Stop Guidelines. Where indented parking is provided, kerb outstand areas at potential bus stops should be long enough to accommodate a future bus stop. Connecting paths and road crossings should be provided during initial development.

ENABLING BUS PRIORITY

The reliability of scheduled bus services can be degraded by congestion, and so priority should be given to public transport vehicles. This is of particular importance at stops, stations, interchanges, major intersections and along key routes and access points identified by PTD. Bus priority must be considered in the context of overall traffic management. Simply increasing road capacity is not the solution. In many instances, the key congestion points will be at activity centres where road space management is a major issue.

Close liaison with VicRoads and the local government authority will be necessary.

Bus priority initiatives include:

- Dedicated bus lanes
- Providing priority lanes where queues may form
- Clearways
- Signal phasing
- Providing bus set backs and lanes at intersections where queues may form eg. queue jump facility
- Introducing and enforcing parking restrictions around and within bus stops.

Fig. Bus priority lane along key routes

Bus priority measures must be in accordance with VicRoads' Bus Priority Guidelines and Austroads' Road- Based Public Transport and High-Occupancy Vehicles.

SMARTBUS

SmartBus is a premium bus service that offers more frequent services and longer operating hours. Buses operate along major arterial roads, connecting activity centres and interchanging with the train and tram networks.

They also provide passenger information displays at selected locations, providing passengers with up-to-the-minute information on bus arrival times. Improved passenger information is also given at every stop with a local area map and service timetable. SmartBus services receive extensive levels of on-road bus priority treatments, such as bus lanes, queue jump facilities and the ability to request green light priority. This helps to ensure a timely and reliable service is provided. Development proposals and traffic management devices should enhance and not prejudice bus operations and travel times. Any delay created must be mitigated to ensure that the service travel time remains unchanged.

TRAMS

Trams are a Melbourne icon and deliver a huge public benefit for Victorians. They are not as operationally flexible as buses and particular planning considerations apply.

Tram services operate both on the road network and in dedicated reservations. Trams running in separate reservations are more efficient, as interaction with private vehicles is more limited. Major objectives for tram services are to improve service reliability by

- Improving speeds
- Removing constraints
- Improving customer accessibility

New urban development should facilitate these outcomes.

Developers should design the interface between the development and the tram network so as to:

- Connect new bus routes to the tram network
- Achieve a high degree of pedestrian and cycle accessibility between the development site and tram stops
- Ensure that tram services are not slowed or otherwise detrimentally impacted.

Early consultation with PTD and VicRoads, Yarra Trams and public transport providers will be required. The tram network must be compliant with the Commonwealth Disability Discrimination Act 1992 and Disability Standards for Accessible Public Transport 2002. PTD is progressively remodelling tram stops with accessible platforms and replacing older trams with new low-floor rolling stock.

The potential for tram extensions will continue to be examined. Developers should liaise with PTD on potential network expansion that may need to be accommodated.

Consultation with tram operators can also be beneficial in providing local level information on routes and operations. The provision of new or revised routes intersecting with the PPTN should reflect consideration of the following:

- Vehicle access points onto and off the tram route
- Right-turning vehicles across the tram tracks including U-turns (especially by taxis), ie. "Left in" and "Left out" only access may be required
- Provision of tram priority
- Maintenance and protection of tram infrastructure, including overhead lines
- Provision of tram stops and passenger waiting facilities including road crossings for pedestrians.
- New route alignments and connections where applicable.

Fig. Tram 109 - recent extension to Box Hill

SERVICE PERFORMANCE REQUIREMENTS

High quality urban design plays an important part in the successful development and operation of the tram system. This includes:

- Enabling tram priority and pedestrian and cycle-friendly traffic management
- Maximising tram catchments through higher density development along routes and around stops
- Minimising right-turning vehicle movements in the path of trams
- Separating trams from traffic where possible.

DESIGN PRINCIPLES (TRAMS)

Infrastructure Design Principles

Location of Stops Along a Tram route

The distance between stops in Melbourne has historically been 200-300 metres. However, there are benefits in increasing spacing to 400-500 metres, eg:

- Average journey times are reduced
- Better facilities can be provided when there are fewer but more accessible stops. An optimal mid block tram stop arrangement is illustrated in Figure.

Some stops in the CAD have been located further apart to achieve these outcomes. PTD and VicRoads have identified a number of priority locations for tram stops in the CAD. Note however, that in high demand areas such as activity centres, more frequent stops may be appropriate. Developers should consult with PTD and VicRoads on how to integrate tram stops into developments, activity centres and the road network.

Accessible Tram Stops

Accessible tram stops fall into two main categories:

- Platform stops
- Kerbside stops.

Platform Stops

Platforms should be at least 33 metres in length, which is sufficient to accommodate the new fleet of articulated trams. A single-facing platform should be 2.4 metres wide. The minimum clear width along the platform to comply with DSAPT 2002 disability access standards is 1.8 metres. Platform facilities such as shelter, audio and real-time information will need to be accommodated. Platform stops can be located at intersections or mid-block between intersections. Mid-block platform stops may also be configured as a central island platform. The central platform width should be 4.5 metres, though high patronage stops may require greater widths for safety.

The key issues to consider include:

- Pedestrian safety
- Urban design
- Traffic capacity impacts
- On-street parking impacts
- Passenger numbers and safe, convenient access to stops
- Laneway access
- Reduced travel times and improved tram reliability.

Kerbside Stops

Kerbside stops are located on the footpath and form the vast majority of stops on the tram network. There are three options to build platforms in these environments: kerb access stops, trafficable platform stops with one traffic lane and trafficable platform stops with two traffic lanes.

Route Upgrade

Development should not preclude the potential to upgrade or extend existing tram routes to provide:

- Traffic light priority treatments at intersections and on approaches
- Increased segregation of trams and motor vehicles
- Separate tram lanes or delineation between tram lanes and other traffic
- Accessible tram stop platforms
- Improved tram stop spacing
- Interchange between other tram services and other transport modes
- Tram stops on the departure side of intersections.

Intersection and Vehicle Access Points

All vehicles except public buses and emergency vehicles are banned from driving along tramways (*i.e.* a section of tram route specifically set aside for trams), crossing a tramway other than at designated locations, and all stopping across tram tracks at crossing points. Tram lanes may be full time or part time and designations may change. Proponents should therefore liase with PTD and VicRoads on the detail of particular proposals. Right-turns may be banned at strategic intersections either all day or during a particular time of day, to ensure trams are not delayed by right-turning vehicles. Where a right-turn must be provided, consideration should be given to fully controlling the right-turn when a tram is detected by the signal.

Vehicles accessing side streets along tram routes often result in trams being delayed. Where tram lanes are not full-time, right-turning vehicles are allowed to enter that lane up to 100 metres prior to turning right. Implementing a right-turn ban at non-signalised intersections, either all day or during a particular time of day, can reduce the interference of vehicles to tram operations. Similarly, U-turns can be banned along particular road sections to reduce disruption to tram operations.

Physical separation standards are being developed for the CAD, the approaches to the CAD, boulevards and commercial/residential urban environments. Physical separation treatments should not exclude bike lanes or overlook pedestrian safety at intersections. The status of these standards should be checked prior to development of designs for physical separation. Double yellow lines along tramways may be used where physical separation is not possible.

Conflict Points for Separated Tram Tracks

In most instances, there is no traffic on separated tram tracks except at intersections and roundabouts.

ENABLING TRAM PRIORITY

Tram priority facilitates improved travel times and increases reliability. The State Government is committed to improving the tram network. Development proposals should take into account measures that will improve tram priority. These measures should also aim to improve passenger and vehicle safety at all intersections, especially those identified as congestion hot spots. The following list of potential treatments and actions may be applicable. The level of priority will be considered by PTD, VicRoads and Yarra Trams on a case-by-case basis:

- Tramways
- Physical separation of tram lanes
- Full-time tram lanes with physical separation
- Part-time tram lanes
- Traffic signal phasing
- Parking bans and full-time tram lanes

- DDA tactile tiles and coloured pavement
- Mid-block parking restrictions to clear tram queues
- Hook turns at signalised intersections
- Paired intersection right-turn operations
- Fully controlled right-turns off tram tracks
- Closure or signalisation of median openings
- Right-turn bans at signalised intersections and facilities at signalised intersections
- U-turn bans along road sections
- Ten-metre long parallel parking bays for drive-in parking
- Review of tram stop locations, eg. relocation of stop to departure side of intersection or mid-block
- Faster stop loading using low floor trams and accessible platforms
- Separation of trams from other traffic at intersections
- Dynamic and/or static signage *i.e.* 'Give Way to Trams'
- Pavement treatment for improved delineation
- Increased visibility of tram clearance line
- Right-turn bans

TRAINS

Melbourne's railway network provides radial public transport linkages to and from the CAD.

Heavy rail services provide the backbone of Melbourne's public transport network and are of increasing importance as the population of Melbourne grows. Trains provide high capacity, fast and direct services, which operate in dedicated corridors with minimal interaction with road traffic. The rail network is a complex system which includes broad and standard gauge lines.

Opportunities to build new rail lines and stations and extend railway lines are being investigated and planning continues for the introduction of new heavy rail services. Developers should liaise with PTD on potential network extensions that may need to be accommodated in the design of the development. In the past, service and network improvements for railways tended to focus largely on capital infrastructure investment. Land use planning should support, complement and build on these investments, so as to optimise land use and transport integration.

SERVICE PERFORMANCE REQUIREMENTS

To ensure the ongoing safety and reliability of existing rail services, a number of key actions are being pursued.

These include:

- Addressing capacity bottlenecks
- Constructing additional tracks
- Upgrading stations to DDA compliance standards

- Improving safety and amenity of pedestrian crossings at both existing and new crossings
- Increasing the number of "Park and Ride" spaces
- Delivering an active urban environment at and around railway stations and interchanges

Proponents are responsible for the interface between a private development and the train network and should consult with PTD. This includes achieving high quality pedestrian and cyclist accessibility between the development and train stations. Proponents also need to consult with key stakeholders, including local government and VicRoads.

DESIGN PRINCIPLES (TRAINS)

Some major developments may propose a new railway station or interchange facilities. Early consultation with PTD will be essential, so that consideration is given to infrastructure requirements, efficient operation and cost. PTD supports proposals which deliver mutual benefits for urban developments and public transport. Each station forms a key node in the public transport network. Stations often act as modal interchanges where the different elements of the public and private transport systems come together and passengers transfer between modes. Key design issues include connectivity, safety and legibility. Specifically these include:

- Grade separation for vehicles and pedestrians
- Provision of clear, continuous and direct pedestrian and cycle routes to transit stops and between modes of transport. Ensure high visibility, activity and surveillance along these routes
- Train facility upgrades or developments along train lines must consider the provision of bicycle and pedestrian facilities
- DDA compliance

Fig. Railway Station Bus Interchange

- Encouraging use of railway station buildings for retail or commercial uses such as cafes, newspaper kiosks and community facilities • Provision of safe, accessible and comfortable bus interchange facilities. These should be integrated into the rail station design to provide maximum ease of transfer for passengers.
- Public transport waiting areas should be clearly visible from the street and adjacent buildings and provide clear views of train, tram or bus arrivals and departures
- Lighting should be well integrated with signage and landscaping in order to maximise safety.

Level Crossings

The design of transport routes at new developments must provide for grade separation at railway crossings except with the approval of the Minister for Public Transport.

Key issues to note include:

- New level crossings are not likely to be approved, ie. it is longstanding policy that at grade crossings should be opposed, except in the most unusual circumstances.
- Development proposals which will generate significant volumes of traffic in the road network around an existing level crossing, should endeavour to direct traffic away from the level crossing by creating alternative access/ egress points.
- Development proposals in the vicinity of existing at-grade crossings will need to reflect any grade separation works proposed for the crossing and/ or respond to the need to upgrade the crossing resulting from demand generated by the development
- Grade separation proposals are likely to impact on access arrangements in and around the crossing and provision will need to be made for alternative access
- Access arrangements will need to consider the requirements of all modes including pedestrian, bus, bicycle, taxi and car.

Car Parks

PTD will determine what constitutes adequate car parking provision based on the type, location and role in the network of each railway station. Proponents contemplating developments close to existing or proposed stations should consult with PTD to ensure adequate consideration of long term parking need and supply. Design of commuter car parking will need to ensure passive security and visual amenity.

Design principles are shown in Figure and include:

- Ensuring car parks are located in areas where active edges provide natural surveillance and enhance the perception of safety for pedestrians.

- Ensuring paths to and from car parks provide appropriate lighting, signage, CCTV surveillance, landscaping and clear sightlines
- Private vehicles should not obstruct access to the station for buses, pedestrians and cyclists
- Providing dedicated pedestrian paths at street level in car parking areas
- Incorporating an appropriate mix of safe bicycle parking at railway stations, which is adequate for local demand
- Decked parking should be encouraged where feasible
- Disabled car parking spaces should be provided at a minimum rate of one in 100 (refer VicRoads Standards).

Short Term Parking and Passenger Drop off

As well as long term car parking, stations require stopping bays for taxis and "kiss and ride" stops. Prominent signage should indicate the short term nature of these parking bays, which may be provided within the car park or on the roadside.

MODAL INTERCHANGES

The quality of modal interchanges is an important determinant of how attractive public transport is and the degree to which it is a viable alternative to private car travel. Modal interchanges range from local suburban bus stops through to major rail interchanges. There are a variety of potential movements to be accommodated at a modal interchange depending on the network connections and services at that site.

A modal interchange may include the following:

- Multiple train, tram and bus services
- Passenger waiting areas
- Pedestrian circulation space for interchange users and non-interchange users
- Bicycle access and storage
- Taxi ranks
- Park and Ride facilities
- Kiss and Ride facilities
- Ferry services.

SERVICE PERFORMANCE REQUIREMENTS

At any interchange, there are a variety of pedestrian and vehicle movements, some of which may conflict with one another. A desirable modal interchange design will enable the efficient movement of interchange users, public transport operators and customers. The following are design principles for an efficient modal interchange.

- Integrate transit stops and interchanges into the design and layout of the activity centre.
- Provide appropriate "Park and Ride" and "Kiss and Ride" facilities in strategic locations.

- Design active frontages along pedestrian paths to interchanges and public transport stops
- Provide direct routes to interchange and ensure high visibility, activity and surveillance along these routes
- Public transport waiting areas should be clearly visible from the street and adjacent buildings and provide clear views of train, tram or bus arrivals and departures
- Lighting should be well integrated with signage and landscaping in order to maximise safety. Lighting should also illuminate timetables at night
- Provide current passenger information about services and the range of service timetables
- Provide directional signage to platforms, stops, conveniences, shops, parking and taxi ranks to minimise confusion
- Additional DDA DSAPT 2002 requirements may be triggered at modal interchanges, such as the provision of resting points (seats) every 60 metres between services.

Liaison with PTD and VicRoads during the pre-application phase is essential. For all development plan and major development proposals, the interchange design must be considered at the preliminary design phase.

DESIGN PRINCIPLES (MODAL INTERCHANGES)

Interchange Layout Design

There are several principal objectives to be met in the design of an interchange layout:

- Maximise passenger and public transport vehicle capacity
- Maximise quality, safety and security of the passenger and operating environment
- Minimise the potential for conflict between passenger, cyclist and vehicle movements
- Minimise walking distances within the interchange and to nearby attractors.

Fig. Transit stops integrated into layout of the activity centre

Fig. Transport Interchange Epping - NSW

There are a number of design issues and constraints which will need to be considered:

- Location requirements:
  - External road layout influences the direction of vehicle flow within the bus interchange
  - Location of trip destinations such as local shops, workplaces, educational institutions, etc, will determine pedestrian 'desire lines' (ie, providing the most direct route reduces unpredictable pedestrian movements)
  - Design of the interchange should allow for direct pedestrian routes whilst carefully locating pedestrian crossing points to maximise safety
- Operational requirements:
  - Efficient movement of trams and buses to and from the interchange
  - Bus layover space will be required at certain interchanges
  - Space for temporary parking or 'stacking' of terminating services
  - Space for originating and through services to pick up Passengers
  - Space for staff facilities and toilets
  - Space for bicycle facilities (eg. bicycle access paths and storage cages)
- Passenger space requirements:
  - For queuing, circulation, seating and any other facilities
  - Passenger information regarding services arriving and departing from the interchange
  - passenger barriers to control movements onto bus services
- Nature of public transport traffic:
  - Number of public transport services per mode, per hour
  - Number of separate routes

- Configuration of buses, as appropriate (eg. Standard or articulated)

Off-Road Interchange Design Options

There are three basic patterns of bus interchange layout and two tram-bus interchange layouts that are preferred, ie.

- Bus-Bus:
  - Islands
  - Concourse
  - Perimeter
- Bus-Tram
  - Cross platform
  - End-to-end

Factors such as available area, site shape and topography, adjoining road layout, relation to pedestrian and passenger objectives, and/or the number of vehicles and passengers will influence the design.

On-Road Bus Interchange

Design and Layout

'Independent' operation is when buses arrive and depart to their own schedule. For example, in most interchanges with multiple bus stops, each bus operates on a different route, to its own timetable. The arrival and departure of one bus should not delay any other bus at a stop in the interchange. For this option, kerb space is required for each bus. 'Nose-to-tail' operation is when buses queue and only the bus at the front of the queue can depart. For example, this could be used when multiple buses operate close together on the same route, or at bus-stops and interchanges with small numbers of boardings and alightings that are not likely to significantly delay following buses. The design (length) of the bays will need to be varied according to whether independent or nose-to-tail operation is required. The following two examples demonstrate successful designs. The choice between parallel bays and sawtooth bays at bus terminals is dependent upon the size and shape of the site and whether independent bus operation is required. (refer to Figures 40 and 41)

- *Parallel:* Design as for on-street parallel bus bays. An additional requirement is for at least a five metre wide adjacent roadway to accommodate buses pulling out from each stop and then passing others which are stationary.
- *Sawtooth:* Sawtooth bays are a variation that permits independent operation within a reduced space. Generally, they are not suitable for public streets because bus drivers may have blind spots behind the angled bus that affect the ability to see approaching cars.

Fig. Parallel bus bays

Fig. Sawtooth bus bays

There are a number of interchange sites, particularly at railway stations, where there is insufficient space for a separate interchange and on-road bus bays will be the preferred solution to bus-rail interchange.

Designs that may direct errant buses towards areas occupied by pedestrians should incorporate physical protection measures.

Taxis

Taxis and mini buses are a flexible form of public transport to meet particular local needs. As a number of taxis are wheelchair accessible, all taxi stops must be DDA compliant.

Major development proposals will need to consider the role and location of taxis and taxi ranks.

Taxi rank design should reflect the following requirements:

- Ranks should be in close proximity to public transport stops to facilitate transfer between modes

- Queuing should be arranged so that people are directed to the vehicle at the head of the rank
- Passenger approach and entry to taxis should be from the passenger side and have satisfactory lighting and signage (*e.g.* taxis should not be positioned along the opposite kerbside from approach direction, as this would encourage passengers to enter while on the roadway)
- If located at shopping centres, clubs and other venues, ranks should be easily accessed without having to cross busy roads, especially by the elderly and disabled persons or substance affected passengers
- Special arrangements may be needed for wheelchairs, as most wheelchair accessible taxis are rear entry and require additional space for this purpose. Wheelchair loading is also time consuming and can delay taxi flowthrough when loading takes place in the rank queue
- Ranks should be positioned so that they do not impede traffic flow
- Ranks should be positioned to the rear of any bus stops to ensure queuing taxis do not block the path of buses.

Further information can be obtained from the Victorian Taxi Directorate, DOT. Also refer to VicRoads' Taxi Rank Guidelines for more information.

Fig. Easily accessibfle taxi ranks

Passenger and Staff Facilities

It is expected that public transport interchange areas will have a range of passenger and staff facilities. The passenger waiting area should include shelter, waiting space, CCTV, toilets, telephones, refreshment facilities, route information, timetables, ticket sales and an enquiry office. In some circumstances, lounge facilities will be appropriate.

Passenger Waiting Area

- Fixed seating should be provided in waiting areas as close as possible to and in view of boarding points

- Seating should be located in a way that does not obstruct queuing passengers.
- Litter bins should be sited close to bus stops, seating areas and main pedestrian routes (provided that arrangements are made for regular removal of rubbish)
- Clocks are of great assistance to operators and passengers
- Glazed screens around the passenger waiting area provide weather protection while enabling waiting passengers to identify approaching buses.
- Toughened glass of between 7.2 and 10.3 millimetres thickness and at least 0.45 metres height should be employed
- Provision for future installation of real-time information systems should be made at all new or refurbished interchanges
- Provision of Metlink signage is mandatory.

Fig. Passenger waiting facilities

Toilets

It is desirable to include passenger and staff toilet facilities at modal interchanges. Provision of public toilets is dependent on the predicted passenger numbers, the availability and proximity of existing public conveniences and the willingness of the relevant local authority to contribute towards operations and maintenance. Where public amenities are provided, it is mandatory to provide a toilet accessible to wheelchair users.

Toilets may be provided in the following combinations:

- Separate staff/ public toilet
- Combined staff/ public toilet
- Staff only – operational staff and interchange staff
- Interchange staff only.

Retail Outlets

Retail uses should be incorporated, as they enhance security, meet customer needs and make the interchange a more vibrant place. A range of uses can be provided, including newsagents, snack bars, etc.

Development Orientation

Developments proposed adjacent to or near an interchange should aim to facilitate

- Active frontages
- Direct and short paths
- Rear parking
- Legible connections to public transport.

10

Rural and Urban Low-Capacity Transport Services

INTRODUCTION

One of the puzzles in the transport sector is that Africa in general, with only a few isolated exceptions, suffers from what has been called the 'missing middle'. Movement of people and goods goes from walking and headloading to the truck and bus in one technological leap. There is nothing much in between. In comparison with Asia there has been little use of animal, bicycle, or motorcycle-based technologies. *i.e.* the sector lacks flexibility in service provision. This has been equally true in urban and rural areas, although the situation is starting to change with the development of passenger carrying motorcycle services in West Africa.

Boda boda transport services are a Ugandan innovation that has grown from small beginnings in the 1960s in the border region with Kenya. The term itself is a corruption of the English 'border border'. Boda boda mainly provide a passenger taxi service, although they can sometimes be hired to move goods. The original services were provided on a man's bicycle, equipped with a padded cushion fitted over the rear carrier. Starting in the early 1990s the bicycle-based carriers have been complemented by, and compete with, light motorcycles that have greatly extended the range and load carriage of services.

The purpose of this article is to describe the nature of the boda boda phenomena. To briefly recap its history, the main lines of development, regulation and organisation, characteristics of its operations, operators and users, and the main problems the industry faces. This analysis is used to examine what benefits the evolution of boda boda services have brought to the poor and to draw some conclusions. It is based on household surveys, and interviews with key stakeholders, operators and users. The studies covered both urban and rural areas, and are part of a DFID-financed research project into 'Sustainable Livelihoods, Mobility and Access Needs (SLAM)'. Household and stakeholder surveys were conducted in four locations in the Kampala – Jinja corridor and these comprised the two cities, a peri-urban location close to Kampala, and a village close to the middle of the corridor. Within each location interviews were conducted with a sample of 90 households, 30 each from poor, medium and high-income groups. Interviews were also conducted with a total of 110 operators and 76 users.

HISTORY

Origins

Boda boda services are known to have originated in the Busia County of Tororo District in Eastern Uganda in the mid-1960s. Confusingly both bicycle and motorcycle services are often known by the same name boda boda, although machala (Western Uganda) or zabala (Mukono District) are preferred in some areas for the motorcycle services. In this paper we will use the term boda boda for both bicycle-based and motorcycle services since it is the most commonly used term. The various 'associations' that have emerged to represent the interests of both groups, often within the same association, also use this name.

Both vehicles are unusual in providing a short-distance, low-capacity service that is able to serve low-density demands, or those where access is restricted by the width or quality of the route. Share taxis do not find it profitable to go to many villages due to insufficient demand. Equally they physically cannot use the footpaths and tracks that provide access to many low-income urban settlements. They operate from 'stands' in towns, in trading centres, and at the bulk public passenger service vehicle stops along main roads that provide access to feeder routes. On these routes boda boda are the dominant service in many areas of the country, both rural and urban. In congested conditions the motorcycles are also valued for their ability to weave through the traffic and shorten door-to-door journey times. Middle-income users tell you they use boda boda when 'in a hurry'.

In isolated areas they may be the only alternative to walking. It has been suggested that in this respect their absence from some villages may be due to the lack of repair facilities and the capital and technical knowledge to set them up. Boda boda also have a not insignificant goods carrying function in the industrial parts of the main cities, and at harvest time in rural areas. One estimate is that this function comprises 30 per cent of urban and rural operations.

It is evident that both services fulfil important economic and social functions, yet little is known about their operations, or those who provide and use them. Hardly a day passes without Uganda's local newspapers having a story or letters related to the boda boda industry. Few seem to be enamoured with them and they have many detractors, but most would concede that the services are indispensable. They deserve the appellation industry because of their scale, and the backward and forward linkages they have with other sectors of Uganda's economy. Although the two vehicle services have separate characteristics there are also strong complementarities.

The introduction of motorcycle-based services is widely reported to have resulted from the initiative of a local firm, BMK (Uganda) Ltd. An interview with the firm's General Manager established that they have been vehicle and spare parts importers and retailers since 1986. A trade visit to Cyprus in 1992 exposed the owner to the use by local farmers of motorcycles. This example, and the knowledge that Japan disposed of large numbers of second-hand and reconditioned motorcycles, led to their introduction into Uganda in the same year. BMK was alone in the market until 1994 when several other firms followed. It was at this time that the phenomenon appears to have taken off as indicated by the figures for newly registered vehicles in Table1.

There are now thought to be about 35 firms, large and small, in the motorcycle import trade. Most bring in 20 (60-65 bikes) or 40 (130-132 bikes) foot containers, but some simply fill the back of a truck (20-30 bikes) that they are also importing.

Fleet Growth

The size and growth of the national motor vehicle fleet in Uganda are subjects of considerable uncertainty due to inconsistencies in official statistics. However, it is clear that there was rapid expansion in the first two thirds of the 1990's, with a substantial slowdown after the end of 1996. Motorcycles appear to have increased in number by a factor of 14, more than double that of the nearest motorised vehicle (pickups and 4-WD). However, the number of newly registered motorcycles has declined by almost 50 per cent since 1997. Since 1999/2000 the growth in all vehicle types has slackened further under the influence of adverse external shocks - an increase in oil prices, the fall in the world coffee price and stagnating export receipts, drought in some parts of the country, and the European Union's ban on fish imports from Uganda, now rescinded.

A contributory factor to the rapid growth of the industry was ease of entry to the market. Government restrictions on the import of vehicles more than five years old were relaxed in 1991. A recent sample analysis of the Ministry of Finance vehicle database, established in 1994, shows that the import of used vehicles is now the principal means of supplying the Ugandan market. In 1999, some 90 per cent of newly registered vehicles were secondhand. Of the 10 per cent of new vehicles, the government imported half. Over 80% of the newly registered cars, vans, pick-ups and mini-buses were declared as used. With motorcycles for commercial use it is thought, by informed observers, that almost 100% are secondhand, although this is likely to change if new Chinese-made vehicles are able to capture market share. The only vehicle category where new vehicles continue to be in the majority is large capacity buses.

The relative cheapness of used vehicles obviously increases the number of new owners, since it lowers the price of entry. It is also easier for existing owners to add to their fleets. The downside of the used vehicle policy is the comparatively high average age (cars 9.4, pickups 6.9, minibuses 7.6, large buses 3.7, and trucks 9.6 years, respectively) of most vehicle categories, so operating costs are also relatively high, and spare parts are a major problem. They might also be expected to have a poorer safety record than newer vehicles, but there is no evidence to show if this is, or is not, the case. It has not been possible to obtain any indication of the growth in the national bicycle fleet. However, the evident increase in all types of boda boda services was clearly the result of both supply and demand factors. A large pool of unemployed, poor and youthful males have provided a ready supply of operators for a highly labour-intensive and arduous occupation.

REGULATION AND ORGANISATION

It has proved difficult for the government to regulate the boda boda industry. However, some operational discipline is provided by the 'associations', to which the majority of boda boda operators belong, although most have only a local jurisdiction. Attempts to form a national organization have been defeated by the chronic instability that plagues most associations. Officers of the current, Kampala-based Uganda Association of Motorcycle & Bicycle Operators (UAMBO) took office only on 15th January 2001. However, they admit to national ambitions and associations have been formed in Bushenyi (Kabwoke), Entebbe, Fort Portal (Kaberole), Gulu, Iganga, Jinja, Luwero, Mbarara and Mpigi.

ROLE OF THE ASSOCIATIONS

Members of an association pay an annual entry fee in the range 6000-10,000 Ushs (US$3.3-5.6). The association acts as their mouthpiece, represents them in cases of harassment by security personnel, traces members in cases of theft, or their relatives if there is an accident, and through their 'stage committees' enforce discipline and hygiene through fines, suspension of membership and the right to operate, and other sanctions. Some have also tried to ensure that members wear a uniform, helmets and ride with proper shoes – slippers are not allowed – and have trained educators for some of these purposes. However, compliance has been mixed, mainly for cost reasons. In some cases the association also has a semi-banking (savings) and credit role for members.

There is some evidence that these associations are changing from their original character of a welfare service for members, to important business and political entities. This has already happened in the taxi industry where the main purpose of their associations now is to collect money from the industry, on behalf of the civic authorities, rather than to put resources into its development. Taxi and bus operator associations not only (loosely) set passenger fares and small goods tariffs, but also allocate, regulate and distribute routes among members. However, it is not known by the author if the boda boda associations exercise a similar function, although anecdotal evidence suggests they do.

The district authorities have the legal right to invite annual tenders from taxi or boda boda 'associations' to collect fees on their behalf from association members. These fees are for the right to use municipal parking facilities and offer services to passengers. Such fees are usually significant contributions to municipal finances accounting for a reported 70% of council revenues in Jinja's case. It is widely believed that the associations are more efficient at revenue collection than the civic authorities, which have a history of fund leakage. Because fee collection is a 'cash' business, and the sums of money involved can be large, especially for taxis, control of the main urban associations has become strongly contested and politicised. Also, a largely uneducated labour force cannot cope with either the technical or political complexity of the tendering process. Splits in the associations and takeover bids are common in what has been characterized by insiders as a mafia business, in which there are strong monopolistic tendencies.

Currently it is unclear whether the operators or the public gain from the current tendering process. A recurrent contemporary issue is the complaint from operator associations that they perceive few benefits from the large sums they hand over to government. From the users perspective it is equally unclear that the industry is truly competitive and provides services at the lowest possible cost. The dominance of the revenue collection role is presently less evident for boda boda associations than taxis, because their turnover is of a different and much lower order of magnitude. They still appear to function primarily as unions, although some of the main urban associations (Gulu, Lira) are collecting revenue on behalf of municipal authorities.

There are boda boda associations operating in Kampala, Mukono and Jinja in the SLAM study corridor. They represent both cyclists and motorcyclists; the proportions of each depending on the factors noted previously - terrain, trip distances and traffic levels. The Uganda Boda Boda Operators Development Association, based in Jinja has some 1800 cycle and 100 motorcycle members. They used to collect fees on behalf of the municipality, but lost the tender to an individual who has reportedly failed to collect efficiently, so they hope to win it back shortly.

OPERATIONS

Since their origin in the Eastern border area of Uganda, bicycle boda boda have spread to most areas of Uganda and parts of Kenya where the terrain permits operations. They operate in only a few flat parts of the capital Kampala, in competition with motorcycles, but traffic and terrain bar them from most of the area. In urban areas where the terrain is flat - such as Gulu, Jinja and Lira - they predominate and far outnumber motorcycles, mainly due to the cheaper fares they are able to offer and the generally short trip distances that appear to dominate.

OWNERSHIP

There are clear differences in the ownership patterns of bicycle and motorcycle boda boda. Most bicycle boda boda (71 per cent) are operated by the owner, compared to 44 per cent of motorcycles. A further 13 per cent of bicycles belong to the family or close relative, and a similar proportion of motorcycles. Correspondingly, hiring of bicycles for operation comprises about 16 per cent of the total, but 56 per cent of motorcycles. This suggests that motorcycle hire is a profitable business that is entered into by people who are already engaged in other economic activities.

It is not obvious why there is a greater propensity to hire out motorcycles. Unfortunately the data on operating costs and earnings are not sufficiently robust to establish if they provide a better return on investment than bicycles. It is possibly significant that few owners had more than one vehicle suggesting that it may not be regarded as a highly profitable industry. However, other explanations are plausible. The scale of the industry and its rapid expansion offers a counter indication that individuals find it profitable. Also, the bicycle sub-sector has a much lower entry cost.

Type of Machines

The motorcycles range in size from 50, to 80, 125 and 250 cc engine capacity. Operators show a preference for the smaller sizes, especially the 50cc Yamaha MATE model, over those with a larger engine capacity. The main considerations seem to be purchase cost - a 50cc motorcycle costs 1.4 million Ushs (US$820) and one of 80cc 1.7 Million Ushs (US$1000) – and fuel economy. The smaller vehicle will enable operation for 10 hours for about 5000-6000 Ushs (US$2.9-3.5). The annual licence fee is also charged at 145 Ushs (US$0.085) per cc per 12 months (*i.e.* US$4.27 for a 50cc motorcycle), so operating costs must be higher for the larger motorcycles. It is not clear if these can be passed onto passengers. Such a small engine clearly limits the payload and range of operations. This must be especially the case in the hilly rural areas that characterise the study corridor, but more evidence on this aspect is required.

Larger off-road motorcycles, up to 250cc, are used in the western border regions because they make long distance trips into the Congo with goods. Similar vehicles were used for transporting fish until the EU imposed an export ban, since lifted. The two largest importers estimated that the bigger goods-carrying motorcycles comprised between 6-8% and 20 per cent of the total motorcycle market. A recent development has been the introduction of new 'Jiange' brand motorcycles from China. Market penetration has been slow due to client conservatism (strong local brand loyalty) and slightly higher prices than for the used motorcycles. However, some of the large fleet clients are showing a preference for new vehicle purchases because of their reliability and assured spare parts supply. Their superior operating costs ought to give them a marketing edge over the second-hand machines, but this is only likely if efforts are made to change the 'economic' perceptions of operators.

There are also reported to be plans to open an assembly plant for Indian-made motorcycles in the 50 – 75cc range. The aim would be to undercut current import prices, but they would also face the problems of brand loyalty, operator fears about their weakness, and their distorted view of the economics of operating old rather than new machines. Most bicycles are imported from China and India, with a few – principally the highly regarded Raleigh brand – from Europe. They are invariably the popular 'roadster ' models dating in design to the 1930's. They cost around 90,000 Ushs (US$50).

Types of Service

Boda boda primarily provide three types of - predominantly - short-distance services: (i) within the main urban areas, where they compete with conventional sole hire taxis and taxis; (ii) as feeders to urban areas on routes that - due either to the low density of demand or the roughness of the route - are unattractive to taxis; and (iii) as feeders to the main roads in which role they tend to complement taxi and large capacity bus services.

The area of operation of a boda boda is called a 'stage'. Each one has a 'stage master' appointed by the association, or selected by the members providing services on that stage, who enforces their regulations. A village might have 2-3 stages with a minimum of 10 registered operators. In contrast in Kampala central alone, there are 124 stage masters and 2000 operators. UAMBO organizes meetings of these stage masters once a month to try to regulate and co-ordinate services.

At each stage there is a committee, headed by the stage master, to instill discipline into operators. If the committee cannot handle a particular case it is referred to the association headquarters.

Areas Served

Bicycle boda boda are restricted to favourable (flat) terrain, comparatively short distances (normally < 5kms) and where the volume of motorised traffic does not militate against their use. They are also more common in the poorer rural areas and small towns. Conversely motorcycles feed off areas where businesses concentrate or incomes are boosted by cash crop production. Thus, bicycle boda boda dominate in the lakeside plain of Jinja (in the ratio 18:1) and the motorcycle version in Kampala (in the ratio 2:1 in the center), which is a city of hills. In the smaller urban areas terrain seems to be the main factor dictating either bicycle or motorcycle use. Gulu and Lira municipalities are both located in the flat northern part of Uganda and are reported to have 1200 bicycle and 60-70 motorcycle, and 3500 bicycle and 20-25 motorcycle boda boda, respectively.

Bicycles are preferred for their cheaper fares, but may not be able to cope either with the terrain or trip lengths demanded. Motorcycles are also said to have a greater accompanied load carriage capacity - a passenger and goods can be carried on the long pillion seat - that in some cases would otherwise necessitate two bicycles. Similarly they have been observed carrying a passenger load of an adult and two children, or three schoolchildren. Bicycle boda boda are also contracted to carry school children, the normal load being two.

Motorcycles generally travel much further than bicycles. When motorcycle operators were asked the furthest distance traveled, the median response was 9 kms, with 37 per cent exceeding 10 kms, and 10 per cent more than 30 kms. The furthest trip was 80 kms. However, pricing for longer trips appears to be opportunistic as there is no tapering of fares per km with distance. This is true even within their main range of operations up to 10 kms, where fares per km tend, as with sole hire taxis, to increase with distance.

Some motorcycles provide classic feeder services to remote villages from the main roads where taxis, large capacity buses or trucks operate. They are also preferred in the rainy season when bicycles are difficult to operate, and at night due to the greater security their speed provides.

Scale of Operations

It is very difficult to estimate the scale of boda boda operations, more so in the case of bicycles than motorcycles. At a national level it is estimated that the proportional breakdown of total motorized travel (in veh-kms) among the different motor vehicle classes is 11 per cent motorcycles, 48 per cent private vehicles, 20 per cent public transport, and 21 per cent commercial vehicles. Since most of the motorcycles operate as boda boda it is clear they make a significant contribution to fare-paying passenger movement.

The total fleet of bicycles can be estimated from the Ugandan national household survey, which records the percentage of households owning a bicycle in different regions of the country. This source yields a figure of 1.7 million. These are used to provide personal mobility, for dedicated rural goods transport, and boda boda services, however, the split between these three functions is not known. Busia District in Kenya, which adjoins Uganda, has about 4000 registered bicycle boda boda. Applying this estimate to districts in Uganda suggests a fleet of about 200,000 vehicles.

Two other estimates give surprisingly similar figures. Surveys in Tororo in 1991 gave a ratio of 1 bicycle boda boda per100 population. This yields a 2001 estimate of 214,000. Secondly moving observer estimates were made by the author in November 2001 of the use of bicycles on the route from Kampala to Jinja, the return journey the following day and from Kampala to Entebbe. A sample of 1100 bicycles indicated that about 12 per cent were used as boda boda, 34 per cent for goods movement, and 54 per cent for personal use. Applying this proportion to the national total of 1.7 million gives a boda boda estimate of 204,000. Although each of the three estimating methods is relatively crude they are sensibly independent, so their convergence on a population figure of 200,000 seems more than coincidental.

The motorcycle fleet can be estimated from new vehicle registration statistics (Table1). Between 1994, thought to be the point of service take-off, and 2000 some 90,000 motorcycles were imported. The overwhelming majority of these are operated commercially and – allowing for scrapping due to old age and accidents, and pre-1994 baseline import levels to reflect non-commercial operations – then the current national fleet of motorcycle boda boda is about 70,000.

Fares

To date there has been only limited investigation of fares in the study corridor. Studies in Mpigi District in 1999, a small part of which is bisected by the study corridor, give some insights. These findings were crosschecked by SLAM spot enquires in February 2001.

- Charges are normally levied for a particular stage, or distance; however, there may be the same charge for stages varying slightly in length.
- As would be expected motorcycles charge higher fares than bicycles for the same stage, but the difference varies with the distance travelled. The average per passenger differences recorded in 1999 were as shown below:

Km	Motorcycle	Bicycle
0-1	500 Ushs (US$0.5)	200 Ushs (US$0.2)
2-4	1000–1500 Ushs (US$1 –1.5)	500-800 Ushs (US$0.5-0.8)

- A charge of 5000 Ushs (US$5) was quoted for nighttime trips emphasising the perceived danger that such journeys pose to operators.
- The foregoing figures all pertain to the then prevailing exchange rate of 1US$ = 1000 Ushs. The exchange rate at the time of the February 2001 enquiries was 1US$ = 1800 Ushs, and it is not clear if boda boda fares have adjusted fully to this, since much lower figures were obtained. The remaining figures all refer to 2001 costs derived from SLAM surveys.
- Currently a typical short stage differential would be 500 Ushs (US$0.28) by motorcycle and 300 Ushs (US$0.17) by bicycle.
- Bicycle operators will accept shorter distance stages (as little as 100 Ushs (US$ 0.06)) than motorcycles.

- Motorcycle costs per km vary in the range 125 Ushs (US$0.07) to 210 Ushs (US$ 0.12) with the lower rates applying to the longer trips up to 16 kms. Comparable inter-urban bus fares per passenger are around 31 Ushs on paved routes and 68 Ushs on unpaved. Thus, motorcycle boda bodas are 2-7 times more expensive than bus fares. However, they are cheaper than sole hire taxi services, which realistically are their most likely competitors in cities.
- The time sensitivity of passengers, shown by their willingness to pay higher fares for motorcycle than bicycle services, is also reflected in the fares charged by buses. The small, but fast, 14-seat taxi mini-bus charges a 500-1000 Ushs premium per stage over their slower 32-seat rivals. Thus, along the entire corridor route Kampala – Jinja a bus operator will charge Ushs 1500 (US$0.83) whereas the faster taxi operator charge Ushs 2500 (US $1.39).

Passengers Carried and Earnings

The number of passengers is very dependent on the location of the stage and competition, with the relatively wealthy city centers generating proportionately more, but shorter, trips than small towns or rural feeder routes that have fewer but longer journeys. Also, not all operators work continuously – many take time off for other activities and occupations. The surveys of bicycle boda boda operators in Jinja found them carrying 35-160 passengers per week, with a mean of 86, or 12 per day. This is the same number per day as was reported for a whole week in Tororo in 1991. The explanation for the apparent discrepancy is that in the latter case trip lengths were at least five times longer than the typical trip in Jinja. In Kampala bicycle operators claimed substantially higher numbers of passengers than in the secondary city Jinja, in the range 75-330 per week, with a mean of 153. In both cities the claimed passengers correlate closely (r2 = 0.86) with stated earnings, which gives credence to the figures. These average 42,300 shs (US$24) per week within a range of 12,000 – 107,000 shs (US$7-59).

Motorcycle boda boda exhibit a similar phenomenon to bicycles with the average number of city center passengers (64) being double those in the peri-urban area of Seeta (32). In Kampala motorcycle operators have a similar range to bicycle operators in Jinja, 18-160 passengers per week, but at 64 the average number is a quarter less than bicycle operators. Again trip lengths normally exceed substantially those of bicycles and fares are higher, factors that are likely to depress the number of passengers. Unlike bicycles the claimed number of passengers of motorcycles operators correlates weakly (r2 = 0.18) with stated earnings suggesting that one or the other, or both, are systematically in error.

Stated earnings for motorcycle operators differ for owners and hirers, and by location. Because of small sample sizes not too much significance can be read into the differences. Owner earnings in Kampala (100,400 shs per week or US$56) are almost double those of hirers (54,200 shs or US$30), but they do have additional costs, especially vehicle depreciation and major repairs. Peri-urban areas yield substantially lower earnings for owners (54,200 shs per week or US$30), but, contrary to logic, hirers appear to earn more (59,200 shs or US$33) per week.

OPERATORS

Operation of the services is an exclusively male preserve. The strenuous nature of the occupation perhaps justifies this with bicycle-based services, but there is no obvious reason why women should not operate motorcycles other than custom and culture, although the long working hours away from home must also be a deterrent. The harassment that operators complain of from some customers and their professed and very real fear of being robbed may be ancillary reasons that deter women from operating such services.

Age

The often-alleged 'youth' of boda boda operators is partially borne out by the SLAM survey data (2001). As might be expected bicycle operators are more youthful than their motorcycle counterparts. The physical demands of cycling favours youth and 20% were 19 or less, and 64 per cent 24 years of age or less. There were none over 34 years old. Motorcycle operators are older with none less than 19 years, only 23 per cent less than 24, and 31 per cent more than 34 years old. About 14 per cent were 40 or more years old. The pooled bicycle and motorcycle operator age distribution is similar to that found in previous studies in Mpigi district, as the figures in Table show. The Mpigi sample may have been somewhat 'older' than average, although none were found over 40 years of age, as it contained retrenched army personnel. They were sufficient in number to have formed their own association – the Bwaise Veterans Association of Mpigi district.

Table. Combined Age Distributions of Bicycle and Motorcycle Boda boda Operators Survey %

Age range (years)	Mpigi	Kampala – Jinja
15-19	8	8
20-29	53	62
30-39	28	21

EDUCATION

A boda boda operator has the popular image of a poor person's 'school dropout' job, which is supported by the description in Annex A and survey data. Some 56 per cent had only primary and 40 per cent secondary education to O level, with just 4 per cent at A and Diploma levels – there were no graduates.

Occupational Significance

83 per cent of bicycle and 80 per cent of motorcycle operators view the provision of boda boda services as their main occupation. The remainder combine vehicle operation with a variety of other income earning activities *e.g.* farming, selling alcohol or second-hand clothes, collecting water, houseboy, brick making, working in a shop, mechanic, and preparing and selling foodstuffs. The apparent occupational dependence is reinforced by the claims of 98 per cent of cyclists and 88 per cent of motorcyclists that it is their main source of income.

MARITAL STATUS AND DEPENDANTS

Among bicycle operators 67 per cent are married, compared to 81 per cent of motorcyclists, probably reflecting the relative 'youth' of the former. The HIV/AIDS pandemic means that many operators support significantly more dependants than might be expected from their comparatively youthful age distribution. As expected dependency is much higher among married than unmarried operators, and slightly higher in both categories for bicycle than motorcycle operators, although it is not clear why.

Table. Number of Dependents of Married and Unmarried Boda Boda Operators

Operators	Number of dependants Married	 Unmarried
Bicycle	7.3	6.2
Motorcycle	1.7	0.5

Allowing for the proportion married and the operators themselves, an average boda boda provides livelihood support to a little over 6 people. Given the previous estimates of the size of the industry these figures imply that about 1.6 million people depend on it for part of their livelihood, which represents about 7 per cent of Uganda's population.

Years Operating

For a variety of reasons operating either type of boda boda appears to be a transient occupation. More than half (52 per cent) had been operating for 2 or fewer years, and only 10 per cent for more than 5 years. The physical stress imposed on cyclists is self-evident, however that on motorcyclists is also considerable with a dangerous operating environment, and violence and theft as ever-present threats. Lack of alternative employment and its profitability probably combine to compel adoption of the occupation.

USERS

In the survey corridor median use of either type of boda boda services by users was 4 years, with 15 per cent claiming more than 8 years and less than 2 per cent more than 10 years of usage. Measurements at the household level show that bicycles account for 8 – 11 per cent of household trips and motorcycles 1 – 3 per cent. Bicycle use, however, tends to decline with income and motorcycle to increase.

OCCUPATION

Interviews showed that users of either type of boda boda are drawn mainly from workers, the business community, students, and health facility patients. People in wage and salaried employment accounted for a little over 43 per cent of users, the self-employed about 38 per cent, with students as an important minority (18 per cent). The latter figure may have been inflated by the results from Kampala with a significant and relatively wealthy University population.

Reasons for Using Motorcycle-based Services

For both men and women it is the speed and convenience of motorcycles that seems to be most prized when: there are no taxis available, a door-to-door service is required, or the user is in a hurry. Some operators are equipped with mobile phones and hence can be summoned. They are also considered to be cheaper than sole hire taxis and faster than bicycles. Load carriage is another important function, especially fish, which taxis will not carry due to the smell. Tiredness, wet weather, terrain and security for women (relative to walking) are minor use factors.

Reasons for Using Bicycle-based Services

The main reasons for using bicycle-based services rather than motorcycle are short distances, to save money and when there is no choice, especially in rural areas. Load carriage, availability, and acceptability of a slower journey are also choice factors. Contrarily lack of a load induces some to take a bicycle rather than a motorcycle.

A significant proportion (25 per cent) does not ever use bicycle boda boda services. A variety of reasons were given: too slow, image and comfort, and fears over safety, being prominent.

INCOMES AND EXPENDITURE ON TRANSPORT

The SLAM household survey results in Table show that within each of the categories there is considerable variation around the mean income due to location. In each category village incomes are substantially the lowest. At a mean of US$50 there is an order of magnitude difference between the incomes of the poor and the high category at US$703 per month. The mean medium category income (US$191) is closer to the poor than high category. Household expenditure on transport is more uniform between locations than income, with Jinja households consistently the lowest, perhaps because it is dominated by (cheap) bicycle boda boda services. The percentage of household expenditure on transport increases broadly with income from a mean of 6.5 per cent to 10.3 per cent. Of particular interest is the extent to which the poorest households are users of boda boda services. This cannot be determined directly, but can be inferred from the data in Table. A village household spends about 1000 shs a month on transport, equivalent to 5 of the shortest (0-1 kms) one-way bicycle stages, or 1 or 2 long (2-4 kms) stages for one person. Alternatively it would buy 2 short one-way motorcycle stages, or perhaps one long stage. Use of boda boda services can thus only be very occasional.

At the other extreme a poor household in Kampala spends 8800 shs a month on transport. Bicycles are rarely available so this is equivalent to 17 of the shortest, or 6-9 of the longest one-way motorcycle stages. Use can thus best be described as intermittent.

So far as income level is concerned the average situation is close to that of Jinja at about 5700 shs a month spent on transport. It would provide at best about 1 one-way short bicycle stage a day, or at worst 4 one-way long motorcycle stages per month. Thus in no location can the poorest households make more than occasional use of boda boda services. The binding constraints are low income combined with the high unit cost of fares.

Lifestyle Changes from Boda Boda Use

For over 90 per cent of men and women users boda boda services have resulted in significant changes in lifestyle by increasing the range and number of the activities in which they engage. They have also enabled them to save time and be more punctual; to make activities easier; and increase their personal performance and output. Only 7-8 per cent perceive no change – they use the services because they are more convenient.

INCOME ENHANCEMENT FROM LIFESTYLE CHANGES

About 60 per cent of male and 38 per cent of female users say that use of boda boda services results in increased income. Prominent among these are traders in bricks, chickens, fish, food, and waragi. They benefit from a greater range of activities, increased volumes transported and sold, and in some cases increased quality through earlier and wider sourcing of produce (fish, fruit, vegetables) - which in some cases has led to increased sales. Traders in cattle, timber and bricks cited their ability to penetrate less accessible areas where they were able to purchase at lower costs than formerly. A number cited the greater mobility of their customers as a factor that had also increased sales.

Others, in wage employment, use boda boda services to gain greater and flexible mobility, and enhance their income by providing more services. For example, supply teachers, using boda boda, can work in several schools, and sales representatives are able to visit more clients. Even apparently static businesses, like shops and hairdressers, claim to have benefited by having more products available for their clients through their ability to source from a wider range of goods using boda boda services. Some users perceive cost savings made by substituting boda boda services for more expensive forms of transport, especially sole hire taxis. Prominent among these beneficiaries are those in salaried employment and university students.

MAIN PROBLEMS

Associations

There is an unfortunate history of instability attached to the associations with recurrent take-overs, financial mis-management and organisational collapse. Local authority officials have formed them from within the industry, but still the instability persists.

In interviews the associations complained that they are weak, with poor management skills, and have little official recognition, and thus influence. The authorities rarely consult them over fee or tax increases. Their control over members is largely voluntary, and indiscipline and reckless driving are thus difficult to regulate. They would, not unexpectedly, like more control over new entrants, which they claim leads to over-competition for passengers.

There is also an evident conflict of interest between their (original) largely welfare and representational function for members, and that of fee collectors for the local authorities. The latter has undoubtedly politicised the work of the associations and led to serious disputes over 'territory' and membership. It has also undermined the original functions of the associations and their ambition to become truly national with a wider range of services to members.

Interviews with association representatives indicated that the main day-to-day issues are:

- Accidents, a common occurrence;
- Theft of vehicles, especially motorcycles;
- Lack of coordination between stage masters, allowing operators to change stages and confuse records;
- Lack of access to micro-finance;
- Poaching of members and their fees;
- Resistance by operators to new ideas *e.g.* helmets and uniforms; and
- Operation of members in areas outside of the association's jurisdiction.

Owners and Operators

Existing and would-be owners and operators face a number of problems including the high cost of entry to the industry, lack of credit facilities, difficulty in obtaining spares, and poor maintenance facilities and skills outside of the main cities. Currently motorcycles attract a combined tax of roughly 55 per cent of the CIF price, comprising 15 per cent import duty, 17 per cent VAT, 4 per cent import commission, 2 per cent withholding tax, and a further

17 per cent corporate tax. Then there is the dealers' profit. Anecdotal evidence suggests that even obtaining a licence can be a considerable financial hurdle for some operators and many do not have licences. Equally it has been suggested that the absence of such services in isolated villages is due to the lack or repair facilities.

Motorcycles are not regarded by the banks as suitable for hire purchase because few would-be owners can provide any collateral or a secure address. The major retailers have in the past offered a limited credit of about 5 per cent of the purchase price and retained the registration papers until the debt has been cleared. However, this has not worked well as some creditors have never returned for the registration papers and BMK (Uganda) Ltd alone has unpaid debts of 200-300 million Ushs (US$0.11-0.17 million) over the past decade. Other retailers say credit default is manageable as long as the Police can ensure reasonable compliance with the vehicle registration regulations by frequent checks that operators carry the appropriate documents.

Trained mechanics and stocks of spares parts are uncommon outside of the main cities. This adds a premium to the cost of rural operations and logically must also lower serviceability levels, thereby further increasing costs or depressing operating returns.

Another frequent complaint is harassment by law enforcement officials and other transport operators, especially taxis. To combat this the association based in the Kenyan border town of Busia has hired its own lawyer to protect its members from abuse of their rights.

In some, mainly relatively wealthy areas, bicycle boda boda have lost market share to their motorised competitors. The poor image and status associated with bicycles are clearly a modal choice factor for some users.

Users

Despite the best efforts of their associations both types of boda boda operators have a poor profile among users, although women (surprisingly) are more charitable, with 28 per cent saying there were no problems. Men complain of reckless, inexperienced driving and distain for traffic rules; dishonesty in overcharging, not having change and actively thieving; poor appearance and personal hygiene; and abusive and arrogant treatment of clients, and coarse behaviour towards women. There is a strong public perception of risk associated with their use either of accidents or disrupted journeys – due to mechanical breakdown or running out of fuel – and even outright connivance of the operators in robbery.

Accidents are certainly common with reckless driving and drunkenness alleged to be the main contributory factors. Youths comprise a significant proportion of boda boda operators and this may be a factor militating against disciplined behaviour. There is some evidence that casualties resulting from accidents involving boda boda occur disproportionately to women. It is not obvious why this should be the case other than their propensity to sit sidesaddle, which implies they are more easily ejected from the vehicles.

Lack of weather protection is also commonly cited as a problem associated with the use of boda boda. While undoubtedly true, given Uganda's tropical rain pattern this is hardly surprising.

BENEFITS TO THE POOR

There are two mechanisms by which the poor can benefit from the operation of both bicycle and motorcycle boda boda: from the employment created, and through using the services provided.

As service providers it has been shown previously that the majority of the operators are drawn from the least educated classes. Among motorcycle operators the low proportion owning their own vehicles, and the high proportion who are married with dependents, also indicate that they are predominantly from the poorest category. It has been shown that 6 people depend for a proportion of their livelihoods on the earnings of a typical boda boda operator. Nationwide the total dependency is estimated at 1.6 million, or about 7 per cent of the population. This excludes those with backward linkages to the industry as mechanics or suppliers of food and drink, etc.

For example, it is estimated that 40-60 boda boda will support a repair shop with 1-2 mechanics. Assuming similar ratios for establishments supplying food and drink suggests that a further 100,000 people might indirectly depend for a proportion of their livelihoods on the boda boda industry by supplying it with repair and sustenance services.

As consumers household expenditure data shows the poorest stratum of the population only occasionally use boda boda services, low incomes and the high unit cost of fares being the binding constraints to increased usage. However, it is not clear to what percentile of the population this applies. Those engaged in formal or informal sector wage employment seem likely to have received benefits from the enhancement of their income earning activities through the greater mobility afforded by boda boda usage.

Isolated communities, urban and especially rural, are special cases. These benefit as boda boda are able to bring in necessities, such as fish, milk, medicines and information.

11

Traffic Engineering

INTRODUCTION

Traffic engineering is that phase of engineering that deals with the planning, geometric design and traffic operations of roads, streets and highways and their networks, terminals, abutting lands and relationships with other modes of transportation for the achievement of safe, efficient and convenient movement of people and goods.

Traffic engineering applies engineering principles that help solve transportation problems by considering the psychology and habits of the transportation system users.

Many people still wonder why a traffic problem is so difficult that an engineer should be called upon for a solution. Why not just install a traffic signal, or raise/lower the speed limit, or erect more signs?

One of the greatest obstacles a professional traffic engineer faces in applying sound principles of traffic engineering is competing theories from vocal non-traffic engineers who lack expertise. The unfortunate result is the creation of traffic hazards when false theories are put into effect. Whenever unnecessary or excessive traffic controls are installed, hazardous traffic conditions usually result.

TRAFFIC SYSTEMS

Traditionally, road improvements have consisted mainly of building additional infrastructure. However, dynamic elements are now being introduced into road traffic management. Dynamic elements have long been used in rail transport. These include sensors to measure traffic flows and automatic, interconnected, guidance systems to manage traffic (for example, traffic signs which open a lane in different directions depending on the time of day). Also, traffic flow and speed sensors are used to detect problems and alert operators, so that the cause of the congestion can be determined, and measures can be taken to minimize delays. These systems are collectively called intelligent transportation systems.

Lane flow equation

Fig. A ramp meter limits the rate at which vehicles can enter the freeway

The relationship between lane flow (*Q*, vehicles per hour), maximum speed (*V*, kilometers per hour) and density (*K*, vehicles per kilometer) is

$$Q = KV$$

Observation on limited access facilities suggests that up to a maximum flow, speed does not decline while density increases. However, above a critical threshold, increased density reduces speed. Additionally, beyond a further threshold, increased density reduces flow as well. Therefore, speeds and lane flows at bottlenecks can be kept high during peak periods by managing traffic density using devices that limit the rate at which vehicles can enter the highway. Ramp meters, signals on entrance ramps that control the rate at which vehicles are allowed to enter the mainline facility, provide this function (at the expense of increased delay for those waiting at the ramps).

HIGHWAY SAFETY

Highway safety engineering is a branch of traffic engineering that deals with reducing the frequency and severity of crashes. It uses physics and vehicle dynamics, as well as road user psychology and human factors engineering, to reduce the influence of factors that contribute to crashes.

A typical traffic safety investigation follows these steps.

- Identify and prioritize investigation locations. Locations are selected by looking for sites with higher than average crash rates, and to address citizen complaints.
- Gather data. This includes obtaining police reports of crashes, observing road user behavior, and collecting information on traffic signs, road surface markings, traffic lights and road geometry.
- Analyze data. Look for collisions patterns or road conditions that may be contributing to the problem.

- Identify possible countermeasures to reduce the severity or frequency of crashes.
  - Evaluate cost/benefit ratios of the alternatives
  - Consider whether a proposed improvement will solve the problem, or cause "crash migration." For example, preventing left turns at one intersection may eliminate left turn crashes at that location, only to increase them a block away.
  - Are any disadvantages of proposed improvements likely to be worse than the problem you are trying to solve?
- Implement improvements.
- Evaluate results. Usually, this occurs some time after the implementation. Have the severity and frequency of crashes been reduced to an acceptable level? If not, return to step 2.

TRAFFIC CHARACTERISTICS

Overview

Information on traffic characteristics is vital in selecting the appropriate geometric features of a roadway. Necessary traffic data includes Traffic Volume, Traffic Speed, and percentage of trucks or other large vehicles.

Traffic Volume

Traffic volume is an important basis for determining what improvements, if any, are required on a highway or street facility. Traffic volumes may be expressed in terms of average daily traffic or design hourly volumes. These volumes may be used to calculate the service flow rate, which is typically used for evaluations of geometric design alternatives.

Average Daily Traffic. Average daily traffic (ADT) represents the total traffic for a year divided by 365, or the average traffic volume per day. Due to seasonal, weekly, daily, or hourly variations, ADT is generally undesirable as a basis for design, particularly for high-volume facilities. ADT should only be used as a design basis for low and moderate volume facilities, where more than two lanes unquestionably are not justified.

Design Hourly Volume. The design hourly volume (DHV) is usually the 30th highest hourly volume for the design year, commonly 20 years from the time of construction completion. For situations involving high seasonal fluctuations in ADT, some adjustment of DHV may be appropriate.

For two-lane rural highways, the DHV is the total traffic in both directions of travel. On highways with more than two lanes (or on two-lane roads where important intersections are encountered or where additional lanes are to be provided later), knowledge of the directional distribution of traffic during the design hour (DDHV) is essential for design. DHV and DDHV may be determined by the application of conversion factors to ADT.

Computation of DHV and DDHV. The per cent of ADT occurring in the design hour (K) may be used to convert ADT to DHV as follows:

$$DHV = (ADT)(K)$$

The percentage of the design hourly volume that is in the predominant direction of travel (D) and K are both considered in converting ADT to DDHV as shown in the following equation:

$$DDHV = (ADT)(K)(D)$$

Directional Distribution (D). Traffic tends to be more equally divided by direction near the center of an urban area or on loop facilities. For other facilities, D factors of 60 to 70 per cent frequently occur.

K Factors. K is the percentage of ADT representing the 30th highest hourly volume in the design year. For typical main rural highways, K-factors generally range from 12 to 18 per cent. For urban facilities, K- factors are typically somewhat lower, ranging from 8 to 12 per cent.

Projected Traffic Volumes. Projected traffic volumes are provided by the Transportation Planning and Programming Division upon request and serve as a basis for design of proposed improvements. For high-volume facilities, a tabulation showing traffic converted to DHV or DDHV will be provided by that Division if specifically requested. Generally, however, projected traffic volume is expressed as ADT with K and D factors provided.

NOTE: If the directional ADT is known for only one direction, total ADT may be computed by multiplying the directional ADT by two for most cases.

Service Flow Rate. A facility should be designed to provide sufficient capacity to accommodate the design traffic volumes (ADT, DHV, DDHV). The necessary capacity of a roadway is initially based on a set of "ideal conditions." These conditions are then adjusted for the "actual conditions" that are predicted to exist on the roadway section. This adjusted capacity is termed *service flow rate* (SF) and is defined as a measure of the maximum flow rate under prevailing conditions. Adjusting for prevailing conditions involves adjusting for variations in the following factors:

- Lane width
- Lateral clearances
- Free-flow speed
- Terrain
- Distribution of vehicle type.

Service flow rate is the traffic parameter most commonly used in capacity and level-of-service (LOS) evaluations. Knowledge of highway capacity and LOS is essential to properly fit a planned highway or street to the requirements of traffic demand. Both capacity and LOS should be evaluated in the following analyses:

- Selection of geometric design for an intersection
- Determining the appropriate type of facility and number of lanes warranted

- Performing ramp merge/diverge analysis
- Performing weaving analysis and subsequent determination of weaving section lengths

All roadway design should reflect proper consideration of capacity and level of service procedures as detailed in the *Highway Capacity Manual.*

THE DEPARTURE CHARACTERISTICS OF TRAFFIC FLOW AT THE SIGNALIZED INTERSECTION

Study on the departure characteristic of traffic flow at the intersection is not only one of the important contents in research on the traffic theories at urban road but also directly affects the setting of signal timing program at the intersection and the optimization of traffic flow organization program. Moreover, it is closely related with the study on the traffic capacity of the intersection. Therefore, profound analysis on the traffic flow characteristics of starting traffic flow, dispersion traffic flow, and stably departure traffic flow in the departure process of traffic flow at the intersection are helpful to improve the level of service at intersection and relieve the traffic jam at the intersection. Based on this, many experts and scholars have conducted extensive researches on the departure characteristics of traffic flow at the intersection and achieved fruitful research results.

For instance, for the micro characteristics of departure vehicles at the intersection, Gazis et al. first proposed stimulus-response model that sensitivity is inversely proportional to time headway on the basis of car-following theory analysis then conducted further researches and deduced the car-following GHR model based on nonlinear algorithm. Through analyzing the micro behavior of traffic flow moving, Yang et al. put forward the concept of random degree of free flow to measure the interactive strength of vehicles in the traffic flow and obtained the traffic dispersion model suitable for the urban roads.

From the aspects of drivers' psychological field features and visual attention features, Jin and Tao analyzed the influence of leading vehicles on the following vehicles in the process of vehicle moving then studied and improved the car-following model for the motion characteristics of following vehicles, respectively. Stokes et al. studied the departure characteristics of left-turn lanes at the intersection, obtained departure saturation flow of left-turn lanes by calculating the saturation time headway, and discovered that there are certain differences in various left-turn lanes. For macro characteristics of departure vehicles at the intersection, May et al. first transformed the traditional research methods, described the formation and dispersion of shock wave at the signalized intersection and bottleneck of pedestrian crosswalk by utilizing graphical method, and then analyzed the signalized traffic flow characteristics and its dispersion process.

Bando et al. studied the overall features of traffic flow from the macro angle, then described the macro features of traffic flow like the unstable phenomenon of traffic flow and blocking process, and obtained the mean-field equation for the relationship between the average speed of the stable traffic flow and the vehicle density. Through analyzing the kinematical characteristics of traffic starting wave at the intersection, Wang et al. defined the significance of kinematics parameter of traffic flow in the traffic wave and built the kinematical model of traffic starting wave at the intersection. In addition, Guo et al. analyzed the dispersion process of traffic flow with graphical method, calibrated the moving speed, motion trajectory, and dispersion time in the departure process of traffic flow on the time-space rectangular coordinate system, and determined the queue length and overall dispersion time of vehicles at the intersection.

To a certain extent, the above research results improve the development of the study on the dispersion characteristics of traffic flow at the intersection. However, it needs further discussion on two aspects. First, the time from the beginning of departure of the traffic flow to stable moving should be measured, as well as the speed of stable moving and the traffic volumes from the departure to passing the stop lines stably. The relationship between the queue length and the platoon dispersion time was analyzed in the above researches, as well as the total time passed through the intersection and passing speed, although it has few discussions to the different stages of departure traffic flow in the intersection, which is difficult to offer accurate based data for signal timing at the intersection and the capacity optimization. Second, the relationships among the spreading speed of the whole traffic wave which produced by the arrived vehicles and the queued vehicles, green time, and the number of the passing vehicles during the queued traffic flow departing are analyzed.

The results presented by the former researchers were only concentrated on the departure process of queued vehicles at the intersection. For the previous studies focus on the release of queuing traffic flow at the intersection, there are few studies on some situations, such as whether the follow-up vehicles could pass through the intersection in the queuing period, the traffic wave propagation situation after the follow-up flow overtakes the queuing flow, and the relationship between the follow-up vehicles and the green time at the intersection and so on. The departure and motion situations of the traffic flow arrived subsequently are closely related to the passing efficiency and signal timing at the intersection. As a result, it needs further research on these aspects.

By summarizing the above research experience, aiming at the traffic flow at the urban typical intersections, this paper applies kinematical equation and traffic wave theory to conduct qualitative analysis for the motion characteristic of the leading vehicle, the motion characteristic of the following vehicles, and the motion characteristic of the whole traffic flow at the intersection during the dispersion process at the signalized intersection. It further builds spatiotemporal model of kinematics in the departure process at the intersection and traffic wave model in the dispersion process at the intersection to determine the changing situations of the leading vehicle at the departure process, the time for the following vehicles to reach the stable speed, and the relationship between the green time at the intersection and the departure vehicle number at the intersection. In addition, in this paper we analyzed the traffic flow departure situations combined with the vehicle operation data at the signalized intersection. Through qualitative analysis and quantitative calculation, we analyzed the dispersion characteristic of traffic flow at the signalized intersection and provided reliable theoretical basis for the setting of traffic signals at the intersection.

MODEL BUILDING

When the traffic lights in the intersection turn green, the queued vehicles start to move. When the space headway between the leading vehicle and the second one is significantly greater than the average space headway of the queued vehicles, it indicates that the traffic flow is beginning to disperse. As the leading vehicle of the queue can accelerate continuously to the expected speed without the leading vehicle ahead, it can easily extend its space headway to the second vehicle. Moreover, the second vehicle will be stimulated to speed up to widen the space headway to the third vehicle. Accordingly, if the conditions are met, the change of the motion state will transfer to the following vehicles continuously so that the whole queued vehicles will participate in the dispersion process. If the green light time is long enough, the transfer process will last until all queued vehicles reach their expected speed by the stimulation of the leading vehicles ahead. Till then, the whole dispersion process of the queued vehicles at the intersection has been finished.

Based on the practical dispersion conditions in the intersection, the paper divides the vehicle motion characteristics during the whole dispersion process into three parts: the motion characteristic of the leading vehicle at the intersection during the departure process, the motion characteristic of the following vehicles at the intersection during the departure process, and the motion characteristic of the whole traffic flow at the intersection during the departure process. Moreover, we build a qualitative description model and make quantitative calculation analysis for these three parts, respectively, so as to analyze the departure characteristics of the traffic flow at urban road intersections accurately and thoroughly.

Assumed Conditions

To simplify the research environment, the following assumptions are proposed.

- There is no bus lane at the intersection or bus stop at the influence area.

- There is no new queue formed at the intersection.
- Take one single lane at the intersection as the research object.
- Ignore the individual differences of vehicles and disturbance of pedestrians and nonmotor vehicles.
- Take the traffic characteristics of straight-moving traffic flow at the intersection with four-phase traffic signal as the analysis object.

Solution Parameters

Based on kinematical equation and traffic wave theory, this paper analyzes the motion characteristics of various traffic flow stages at the departure process in the intersection and solves the following traffic parameters:

- The time from the beginning of departure of the queued vehicles to stable moving,
- The departure vehicle number from the beginning of departure of the queued vehicles to stable moving,
- The time headway when the queued vehicles accelerate and stably move,
- The total time needed in dispersion the queued vehicles,
- The spreading speed of the traffic wave after the confluence of the vehicles arrived subsequently and the queued vehicles,
- The relationship between green light time at the intersection and the number of the departure vehicles at different situations.

TRAFFIC IMPACT ANALYSIS

Understanding the demands placed on the community's transportation network by development is an important dimension of assessing the overall impacts of development. All development generates traffic, and it may generate enough traffic to create congestion and to compel the community to invest more capital into the transportation network, whether it is in the form of new roads or traffic signals or turn lanes. Traffic congestion results in a number of problems, including economic costs due to delayed travel times, air pollution and accidents. As one roadway becomes congested, drivers may use others not necessarily intended for through traffic. As a result, traffic impact analyses are becoming more common as a planning tool to fore-see demands on the transportation network and to mitigate any negative impacts. Understanding traffic impacts becomes even more important as budgets for public facility and infrastructure improvements become increasingly strained.

WHAT IS TRAFFIC IMPACT ANALYSIS (TIA)?

A traffic impact analysis is a study which assesses the effects that a particular development's traffic will have on the transportation network in the community. These studies vary in their range of detail and complexity depending on the type, size and location of the development. Traffic impact studies should accompany developments which have the potential to impact the transportation network. They are important in assisting public agencies in making land use decisions. These studies can be used to help evaluate whether the development is appropriate for a site and what type of transportation improvements may be necessary.

Traffic impact studies help communities to:

- Forecast additional traffic associated with new development, based on accepted practices.
- Determine the improvements that are necessary to accommodate the new development.
- Assist communities in land use decision making.
- Assist in allocating scarce resources to areas which need improvements
- Identify potential problems with the proposed development which may influence the developer's decision to pursue it.
- Allow the community to assess the impacts that a proposed development may have.
- Help to ensure safe and reasonable traffic conditions on streets after the development is complete.
- Reduce the negative impacts created by developments by helping to ensure that the transportation network can accommodate the development.
- Provide direction to community decision makers and developers of expected impacts.
- Protect the substantial community investment in the street system.

Traffic impact analysis is only one component of the larger transportation puzzle. In addition, large communities in particular will need to determine appropriate mixes of transportation modes, including public transit options. Community growth pat-terns and characteristics can be substantially affected by highway expansion or re-alignment decisions made at state or federal levels. Traffic impact analysis is focused on the effects of a particular set of developments, but may provide information relevant to these broader plans and decisions. Traffic impact studies should be used as one piece of several kinds of information to judge the suitability of development from a transportation standpoint.

TRANSPORTATION IMPACT ANALYSIS (TIA) REQUIREMENTS

A Transportation Impact Analysis (TIA) evaluates the adequacy of the existing transportation system to serve a proposed development, and the expected effects of the proposed development on the transportation system. The TIA should provide adequate information for County staff to evaluate the development proposal and, when appropriate, recommend conditions of approval. Throughout the Transportation Impact Analysis process (and beginning as early as possible), cooperation between County staff, the applicant, and the applicant's traffic engineer is encouraged to provide the best possible conditions for the traveling public and potential users of the proposed development, and to reduce TIA report revisions and review time. If County staff can be of assistance in any way during this process, or if any questions arise about this process, please do not hesitate to consult us for clarification or assistance.

Marion County staff may, at its discretion, and depending on the specific situation, require additional study components in a TIA or waive requirements deemedinappropriate. Marion County staff may waive a TIA that would otherwise be required if the developer agrees to certain conditions of development. Marion County assumes no liability for any costs or time delays (either direct or consequential) associated with Traffic Impact Analysis preparation and review. Marion County Public Works reserves the right to charge an hourly fee to cover staff time forexcessive or repeated reviews necessitated by TIA inaccuracies or deficiencies.

Transportation Impact Analysis Be Required

A Transportation Impact Analysis shall be required for:

- Any proposed development that can be reasonably expected to generate more than 600 vehicle trip ends during a single day and/ or more than 100 vehicle trip ends during a single hour.
- Any proposed zone change that, in typical build-out scenarios, can be reasonably expected to generate more than 300 vehicle trip ends more than the previous zoning during a single day.
- Any development within the Urban Growth Boundary of a city if the development would meet that city's criteria for requiring a Transportation Impact Analysis.

A Transportation Impact Analysis may be required for:

- Any proposed development that can be reasonably expected to generate more than 200 vehicle trip ends during a single day or more than 40 vehicle trip ends during a single hour.
- Any case in which, based on the engineering judgment of the Public Works Director, the proposed development or land use action would significantly affect the adjacent transportation system. Examples of such cases include, but are not limited to,: non-single family development in single-family residential areas, proposals adding traffic to or creating known or anticipated safety or neighborhood traffic concerns, or proposals that would generate a high percentage of truck traffic (more than 5 per cent of site traffic).

Calculation Of Trip Generation And Distribution

Trip generation data provided in the most recent edition of the ITE publication *Trip Generation* should be used unless more appropriate data is available. Average trip generation formulas (where applicable) or rates are normally used; however, more conservative calculations may be required by staff in some cases. Directional trip distribution assumptions should be based on historical data, existing and future travel characteristics, and capacity constraints. County staff may require data collection at similar facilities if County staff determines that insufficient trip generation data is currently available. *To reduce revisions and review time, approval of the trip generation and distribution assumptions (including any applicable pass-by, internal, or diverted linked trip percentages) and methodology should be obtained from the Public Works Department before using these assumptions in the Transportation Impact Analysis.*

Determination Of The Area For Which Analysis Is Required

The Transportation Impact Analysis shall address at least the following areas:

- All proposed site access points.
- Any road segment or intersection where the proposed development can be expected to generate more than 360 additional vehicle trips during a single day or more than 60 additional vehicle trips during a single hour (these typical volumes may need to be adjusted for unusual situations, such as heavy truck traffic, safety issues, or capacity limitations). If a two-way-stop controlled intersection currently functions acceptably and the proposed development would be expected to generate a total of less than 60 additional vehicle trips per day on the minor leg(s) of the intersection, it need not be included in the study area as a result of this requirement. County staff may, at their discretion, choose to waive study of certain intersections if they deem such study to be unnecessary.
- Any road segment or intersection where the additional traffic volume created by the proposed development is greater than 10 per cent of the current traffic volume (for road segments) or the current entering volume (for intersections). Public Works staff may, at their discretion, choose to waive study of certain intersections in some cases.
- For developments expected to generate more than 30 truck trips per day, the TIA study area shall include the route(s) that these trucks would take from the site to and from the arterial system.
- Any other intersections adjacent to the subject property.
- For developments expected to generate a significant percentage of truck traffic (more than 5 per cent of site traffic), consult Public Works staff to determine the study area.
- Any other intersections identified by Public Works staff as having capacity, safety, neighborhood, and/or geometric concerns.

Consultation in advance with Public Works staff to determine the extent of the study area is strongly encouraged.

Horizon Year

The horizon year of a Transportation Impact Analysis is the most distant future year that shall be considered in the Transportation Impact Analysis. The horizon year will be a specified number of years after the development opens, and this number will vary depending on the size of the development, any land-use plan changes necessary to allow it, its uses, and the anticipated time until full buildout. The following table shows the TIA horizon year (expressed in years after the development is planned to open) for developments expected to generate less than 5 per cent truck traffic:

Development Type/ Trip Generation Per Day	Horizon Year
Any Zone Change	20 years
Other Development, Less Than 1,000	0 years
Other Development, 1,000 to 1,999	5 years
Other Development, 2,000 to 4,999	10 years
Other Development, 5,000 or more	20 years

For developments expected to generate more than 5 per cent truck traffic, consult County staff for the TIA horizon year. County staff may, at their discretion, reduce the horizon year in cases where less future study is necessary.

Transportation Impact Analysis (TIA) Report Requirements

The preparer is encouraged to coordinate preparation with County staff and staff from other jurisdictions, as appropriate to ensure that all necessary components are included in the TIA and to reduce TIA revision and review time.

In order to be reviewed, the Transportation Impact Analysis (TIA) report shall include at least the following minimum components (incomplete reports will be returned to the applicant's representative for completion):

- The TIA report shall be signed and stamped by a Professional Civil or Traffic Engineer registered in the state of Oregon.
- An executive summary, discussing the development, the major findings of the analysis, and the mitigation measures proposed.
- A vicinity map showing the location of the proposed project in relation to the transportation system of the area.
- A complete description of the proposed development, including a site plan, with the best available information as to the nature and size of each proposed use, and the proposed location and traffic control of all proposed access points (including the distance from all proposed access points to adjacent accesses and/or streets).

- A brief description of the current (and proposed, if applicable) land uses adjacent to the site, including the location, size, zoning, current use, and future use of any land parcels that are not part of the subject application, but may use the subject parcel for all or part of their access. If there is potential for development of these parcels, include the best available information as to the potential future use of each parcel.
- A description of the TIA study area, including roadway names, locations and functional classifications, intersection lane configuration and traffic control (including signal timing), existing Right-of-Way, transit routes and stops (if any), pedestrian and bicycle facilities, and planned transportation system improvements.
- Existing traffic volumes (measured during design conditions and/or the peak season within the previous 12 months, unless County staff deems newer counts necessary due to recent development or seasonal variations). Consult County staff to determine what type of count data (turning movement, ADT, or classification) is necessary.
- Accident data within the study area for the most recent available three year period (accident data can be obtained from the Oregon Department of Transportation).
- Existing performance of the transportation system, including Levels of Service (LOS) and Volume/Capacity ratios (V/C) for all intersections and road segments as appropriate within the study area.
- Complete trip generation figures for all aspects of the proposed development, including number of trips by vehicle type and size, and time-of-day and entering/exiting percentages. These figures shall include trip generation figures for any other proposed developments on the subject property, and/or any proposed developments that would share access with the subject property. For developments expected to generate a significant amount of truck traffic (more than 30 trucks per day), include separate figures for trucks. Document the sources of this trip generation data. If the source is other than ITE's *Trip Generation*, the preparer must obtain approval of the use of such data from County staff before using it in the TIA.
- Trip generation figures for any pending and approved developments that would affect the study area. County staff will facilitate procurement of applicable data in these cases.
- Identification of the critical analysis period(s) and justification of this identification.
- Trip distribution for the proposed development. For developments expected to generate more than 30 truck trips per day, include separate trip distribution figures for trucks.

- Forecast traffic volumes without the development, in the year that the proposed development is planned to open, and in the horizon year (consult County staff for information to determine these future traffic volumes). If phased development is proposed, include projections for the year that each phase of the development is planned to be complete.
- Forecast performance (including LOS and V/C) of the transportation system without the development in the year that each phase is planned to be complete and in the horizon year.
- Forecast traffic volumes, including the proposed development traffic, in the year that each phase of the development is planned to open, and in the horizon year.
- Forecast performance (including LOS and V/C) of the transportation system, with the proposed development, in the years that each phase of the proposed development is planned to open, and in the horizon year. Include analysis of signal warrants, signal progression, queue lengths, and other traffic flow characteristics as appropriate. For developments expected to generate a significant percentage of truck traffic, demonstrate how the analysis adequately accounts for the presence of these trucks in the traffic flow.
- Safety analysis of the site accesses, including sight distance and operational characteristics.
- Analysis of right and left turn lane warrants, queue lengths, acceleration lanes, throat lengths, channelization, and other characteristics of the site accesses as appropriate.
- Comparison of the location and spacing of the proposed accesses with Marion County standards, the standards of the appropriate city for developments within Urban Growth Boundaries, and/or Oregon Department of Transportation standards for developments near state highways.
- Analysis of the parking needs of the proposed development, the adequacy of the proposed facilities to meet those needs as appropriate, and the conformance of the proposed parking facilities to applicable standards.
- Evaluation as appropriate of the turning and traveling characteristics of the vehicles that will be using the proposed development and the adequacy of the geometrics of the existing and proposed roadway (public and/or private) configurations to accommodate these characteristics.
- Analysis as necessary of the adequacy of the internal vehicle and pedestrian circulation systems to serve the proposed development and how the design of the development addresses the Transportation

Planning Rule requirements regarding pedestrian-, bicycle-, and transit-friendly developments.

- Analysis as appropriate of any potential adverse or controversial effects of the proposed development on the transportation system or quality of life in the area. Examples of possible effects include, but are not limited to, infiltration of non-residential traffic into residential neighborhoods, traffic noise, creation of potential for traffic violations, conflicting turning movements with other driveways, etc.
- Analysis as appropriate of the effect of the proposed development on pedestrian and bicycle transportation in the area, and any new pedestrian or bicycle transportation needs arising from the development.
- Listing of all intersections and locations that are projected to not meet Marion County (or other jurisdiction, as appropriate) intersection performance standards in the TIA study area during the required analysis period.
- Description and analysis of mitigation measures necessary to bring these intersections and locations into compliance with the applicable standards. Include analysis showing that these measures will bring these locations into compliance and include signal, turn lane, or other warrant analyses as appropriate. The TIA shall also specify the timing and phasing of any new traffic signals and the length of any new turn lanes. Any mitigation measures recommended in the TIA shall be physically and economically feasible, and this feasibility may need to be demonstrated in questionable cases.
- Copies of raw traffic count data used in the analysis (this may be presented in an appendix).
- Calculation sheets and/or computer software output for all LOS and V/C calculations in the analysis. For signalized intersections, this must include the signal timing used in the analysis (this may be presented in an appendix).
- Warrant worksheets for signals, turn lanes, signal phasing, all-way-stops, and other proposed measures as appropriate (this information may be presented in an appendix).

Additional Study Requirements

The basic TIA report requirements are listed in the previous section. Additional information and analysis will be necessary to properly analyze many development scenarios, and the Transportation Impact Analysis shall include a complete analysis of the existing conditions and the proposed development. The applicant and/or the traffic engineer can and should submit any additional information that may be helpful to County staff in understanding the proposed development and/or the traffic that it would generate. County staff may require additional study beyond the scope of the original TIA, especially in cases where additional transportation system concerns arise either as part of the traffic analysis process, as part of the approval process, or from the general public. County staff may also, at their discretion, choose to waive certain report requirements where they deem such analysis to be unnecessary. Please do not hesitate to contact County staff if there is any question as to whether or not certain analysis information should be included in the TIA.

Methodologies and Analysis Parameters

- All signalized and all-way-stop controlled intersections shall operate at a Level Of Service D or better (all individual movements shall operate at LOS E or better) with a Volume/Capacity ratio of 0.85 or less. Other unsignalized intersections (including unsignalized private accesses) shall operate at Level Of Service E or better, although LOS F may be allowed if the movement has a relatively low volume (as determined by County staff) and there is no indication that a safety problem will be created. Intersections within the Urban Growth Boundary of a city shall also meet the intersection performance standards of that city. Intersections near state highways shall also meet the standards of the Oregon Department of Transportation.
- Acceptable analysis methods include the most recent Highway Capacity Manual, PASSERII, HRR211, TRANSYT-7F, SIGCAP, and UNSIG10 for most cases. For high percentages of truck traffic, unusual types of intersections, or other cases which do not specifically fit the circumstances for which the above analysis tools are intended, or if the engineer believes that another analysis method more accurately models the situation, consult County staff for determination of the appropriate analysis procedure. Analysis performed using methods not accepted by County staff will be returned to the applicant's representative for revision and correction.
- Signal timing used in capacity or progression analysis shall use the same cycle length as is currently in use at the intersection, unless specifically noted otherwise, and shall not exceed 136 seconds. Signal timing shall provide adequate available green time (according to Marion County standards) for pedestrian crossing in all directions, and shall provide a minimum of 15 seconds of available green time for protected left turn phases, and a minimum of 10 seconds of available green time for protected/permissive left turn phases. Current yellow and all-red time shall not be decreased.
- Saturation flow rates greater than 1800 passenger cars per hour per lane shall not be used unless specifically measured at that location.

- Peak Hour Factors greater than 0.85 shall not be used unless justified by specific counts at that location.
- Arrival Type 3 (random arrivals) shall be used in signalized intersection analysis unless specific measurements at that intersection indicate otherwise.
- Signal Progression shall be analyzed in all cases where either a new signal or a change in signal timing is proposed on a roadway with more than two traffic signals (including the new signal, if appropriate) in the space of one mile. A minimum greenband width equal to 40 per cent of the cycle length shall be maintained on all arterials, at a progression speed within five miles per hour of the posted speed limit.
- Any proposed signal timing shall provide adequate green time for pedestrians to cross all legs in all directions, at a speed of 4 feet per second, plus a six-second cushion.
- All calculations and analysis results should be reasonable, understandable, consistent, and fully explained. Calculations, graphs, tables, data, and/or analysis results that are contrary to good common sense will not be accepted, and may lead to the TIA being returned to the applicant's representative for correction.
- The conclusions presented in the TIA shall be consistent with and supported by the data, calculations, and analysis in the report. Inconsistent and/or unsupported conclusions will not be accepted, and may lead to the TIA being returned to the applicant's representative for correction.
- Provide two copies of the Transportation Impact Analysis report for County Staff to review. If any portion of the study area falls within another jurisdiction (such as a state highway or a city), consult that jurisdiction to determine the number of additional copies that they will need for their review.
- The attached checklist will be used by County staff to determine if a TIA contains sufficient information to be reviewed. Incomplete and/ or unacceptable TIAs will be returned to the applicant's representative for completion and/or correction. Acceptance for review does not certify adequacy and is in no way an approval. Additional information may be required after acceptance of the TIA for review.
- Cooperation between the applicant, the applicant's traffic engineer, and County staff is strongly encouraged throughout the TIA process. The applicant or applicant's traffic engineer should not hesitate to contact County staff if any uncertainties should arise.

ANALYSIS OF NEED FOR TRAFFIC MITIGATION

Decision Process

Traffic mitigation decisions are, by necessity, made on a case-by case basis.

Each project is unique. The extent and types of traffic mitigation strategies selected for a project will be determined by the characteristics of the project and the site in which it will be located. Additionally, some projects offer special traffic mitigation challenges and some strategies will be better able than others to accomplish these needs.

Some projects cause little disruption to motorists and will need nothing more than modest signage to inform motorists that construction is ahead. Other projects, especially those for which substantial travel delays are expected, might require numerous TM strategies, including strategies that divert traffic to alternate routes, provide new travel options, inform a wide audience of travellers about the construction, or alter construction procedures.

This section describes the types of project characteristics that can influence TM decisions and offers guidance on how to determine the general level of TM needed for a project and to estimate an order-of-magnitude TM cost. Following this section, guidance is presented for choosing individual TM strategies.

PROJECT CHARACTERISTICS RELEVANT TO TRAFFIC MITIGATION

The number and mix of strategies included in a traffic mitigation program will depend on many factors, including the scope and duration of the project, the volume of traffic currently using the roadway, characteristics of the construction site, the construction activities that must be accomplished, and the travel alternatives available to motorists. A package of strategies must be chosen that will address the specific challenges for that project. Many project characteristics could influence the need for traffic mitigation and the extent and type of traffic mitigation strategies chosen. Several characteristics are considered to be especially important. These include characteristics related to the level and type of impacts the project is likely to cause to motorists and to members of the community in which the project is located. These primary characteristics should be examined early in the TM process

Primary TM Project Characteristics:

- Road User Cost of Staging
- Impacts to retail business
- Impacts to neighborhoods
- Impact on military and emergency response
- Impacts on access to major activity centers or employers
- Impacts on local/developer projects
- Constructability
- Political sensitivity

Other characteristics have a less direct relationship to TM but also might be useful to examine to help identify specific TM strategies that could be useful to implement. These secondary characteristics include the following:

Secondary TM Project Characteristics:

- Duration of construction (months)
- Number of major construction stages
- Lane and ramp closures
- Type of roadway – Interstate, State, major arterial
- Average daily traffic (ADT) volume
- Seasonal recreational traffic volume increase
- Expected delay (vehicle hours per day) or queue length (miles)
- Distance covered by the project
- Level of urbanization in the construction area
- Proportion of trucks in the traffic mix
- Extent of public/media exposure
- Nearby projects – influence and impact
- Maintenance considerations
- Holidays and schools

PRIMARY TM PROJECT CHARACTERISTICS

Road User Cost

Road user cost (RUC) is an economic measure of the impact of construction on motorists. It is calculated as the added vehicle operating costs and delay costs borne by highway users as a result of construction projects. These costs are a function of the timing, duration, scope, and characteristics of the construction project, and the volume and operating characteristics of the traffic affected. Road user cost is important in TM because it helps to define an appropriate expenditure for TM activities. A high RUC suggests a greater investment should be made in TM. But RUC also sets an upper limit for the TM investment. In general, the total cost of TM should not exceed the total RUC.

Impacts to Retail Businesses

Some construction projects will have a direct impact on retail businesses because they may make the businesses inaccessible, reduce parking areas, or divert traffic away from the business's locations. Roadway signage can be used to promote these businesses and indicate how they can be accessed. Construction strategies and some traffic control and operations strategies need to be particularly well-planned when businesses are affected, but strategies that reduce the duration of construction could be very appropriate.

Impacts to Neighborhoods

Construction projects also can create significant problems for residents in neighborhoods near the construction area. As for business, construction can hinder access and reduce parking. Construction strategies and some traffic control and operations strategies that minimize the flow of traffic through the area and/or reduce the duration of construction could be very appropriate.

Impacts on Military and Emergency Response

Construction projects can disrupt the ability to respond to emergencies. This should be considered when developing traffic staging plans, temporary detours, or changing access to residential and business areas.

Impacts on Access to Major Activity Centers or Employers

Construction areas that pass through or near major activity centers or employment areas can create substantial traffic back-ups, especially at peak travel times. Generally, the more commercial centers and employers are affected by construction, the more TM planning should emphasize public information, construction strategies, alternate routes, and Travel Demand Management (TDM). TDM strategies, such as rideshare incentives, transit improvements, and Park & Ride lots should receive emphasis when they can be targeted to major employment centers.

Constructability

Key characteristic which must be considered early on is the ability to construct a project in the manner envisioned. Soliciting input early on from individuals with construction expertise and the ability to consider abstract ideas in three dimensions are necessary to ensure that the project can be built as planned.

Political Sensitivity

Projects with high political sensitivity include those in which lawsuits are underway or threatened, or where political leaders and interest groups are taking highly visible positions either for or against the project. These projects require special emphasis on public information and motorist information. They also might require attention to alternate routes and lane and ramp closures proposed for the construction. Additionally, legislative offices should be kept informed of project progress through legislative liaisons and fax networks.

SECONDARY TM PROJECT CHARACTERISTICS

Duration of Construction (months)

The length of time construction is expected to last has a significant impact on both the extent of TM and the types of strategies chosen. Long projects generally can need and can justify a greater number of strategies and more aggressive strategies than can short-duration projects. Although capital improvements might be justified if they provide permanent improvements, large scale and capital intensive strategies are generally used only for large projects.

Number of Major Construction Stages

A multi-staged project can create frequent changes in the availability of and accessibility to the roadway, leading to motorist confusion. This suggests a greater emphasis on public and motorist information about scheduled lane and ramp closures. Multi-stage projects also might be appropriate for construction strategies, alternate route, and TDM strategies.

Lane and Ramp Closures

The need to close lanes or ramps can increase construction delay significantly. The greater the loss of capacity, the more TM will be needed. In these cases, strong consideration should be given to detours, incident management, and motorist information. Public information about the closures also should be considered to give motorists advance notice of the changes.

Type of Roadway

Interstates and State expressways typically warrant more extensive TM than do other roads because of their high traffic volume and critical function in the roadway network. The need to complete construction quickly while maintaining traffic flow suggests strategies that reduce the duration of construction as well as incident management strategies. Because more of the traffic on these roads comes from beyond local areas, motorist information also is likely to be important.

Average Daily Traffic Volume

The traffic volume (24 hour/7 day) through the corridor has an important influence on the level of TM needed. In general, the higher the volume, the more motorists affected, the greater the total cost of delay, and the greater the justification for TM investment. High volumes suggest the use of incident management, TDM, detours, and public information strategies.

Seasonal Traffic Volume Increase

Some highways experience significant increases in travel during certain seasonal recreational periods. If construction is scheduled on these roads during seasonally high volume times, more extensive traffic mitigation strategies could be required than during the lower volume time period.

Expected Delay (vehicle hours per day and per project duration) or Queue Length (miles)

Delay is a significant TM characteristic because it is a key measure of the degree of disruption to motorists. Experience suggests that some of the least costly TM strategies for reducing delay over the project period include incident management and contract incentive/disincentive strategies.

Project Distance (miles)

The length or breadth of the project can influence the types of traffic mitigation strategies chosen. Projects that extend over a large geographic area might warrant the use of detours and broad-based public information.

Level of Urbanization

The level of urbanization frequently is related to the ADT for the roadway. Projects in highly urban areas generally warrant higher levels of TM than do projects in suburban or rural areas and offer opportunities to implement broad-based public information, such as radio ads; this outreach is effective because the proportion of traffic that is local is usually high, and motorists can be reached relatively easily with print and electronic media.

TDM strategies, such as additional transit and Park & Ride lots, also are more likely to be useful in urban areas where population densities are greater.

Significant Truck Volume

Projects with significant truck volumes can increase the need for TM because they increase vehicle delay. Where truck traffic will be heavy, attention should be focused on construction and detour strategies to divert trucks and on incident management strategies that have the capacity to handle large trucks and load spills. The volume of trucks also can affect the use of and selection of alternate routes.

Public and Media Attention

Some projects will attract considerable public and media attention because of the project location, duration, and expected delay. These projects deserve special attention to public information, motorist information, and incident management.

Nearby Projects

Consideration of construction planned for parallel routes or other projects in the vicinity is needed. The use of common detour routes may be problematic, so too the exponential increase in congestion caused by two projects along the same corridor.

Maintenance Considerations

Snow removal, mowing, and other maintenance duties during construction may complicate projects and detract from the overall traffic mitigation strategy.

Holidays and Schools

A special effort should be made to safeguard school bus routes, school crossings, and other needs of school children. Holiday shopping may impact traffic in the corridor and should be considered in the overall traffic mitigation strategy.

LEVEL OF TRAFFIC MITIGATION APPROPRIATE TO OR INDICATED BY PROJECT CHARACTERISTICS

As a general rule, the extent of expected delay and RUC will be determining factors in the requisite TM level. They are fundamental TM measures because they capture the impacts of other variables, such as the traffic volume, length of construction, and proportion of trucks in the vehicle mix. Frequently, several of the characteristics described above will appear together in a project. For example, long duration projects often are completed in stages and result in extensive cumulative delays; projects on high volume roads often result in long delays; roads in highly urbanized areas generally carry high traffic volumes; and projects with high impacts to businesses/ employment centers might be associated with political sensitivity. When multiple TM challenges exist, the level of TM likely should be elevated.

In order to best assess traffic impacts, accurate and current traffic data is needed. As such, the Division of Project Planning and Development (DPPD) will acquire 24 hour weekday and weekend truck/car counts for the project area. These should include mainline volumes and, if applicable, side street volumes at signalized intersections and detour routes. Turning movements at key intersections and interchanges shall be provided for peak hour periods. This data shall either be created by, or forwarded to, the Bureau of Transportation Data Development (TDD) which shall maintain a centralized data base of all traffic volume counts.

The evaluation of traffic volumes through various work zone capacities and detour routes will help the Designer understand traffic impacts associated with various lane closure alternatives. The Designer will evaluate the alternatives, on a 24 hour basis, to determine when there will be queues.

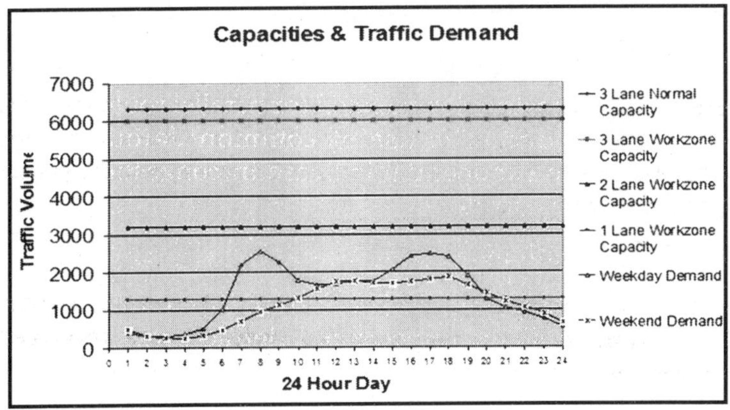

For evaluation of detours, the detour length and delay times shall be provided by the Designer. The NJDOT Road User Cost Manual provides greater detail in these matters. The number of lanes available in a work zone plays a significant role in the work zone capacity. Work zone capacity charts are in the NJDOT Road User Cost Manual. Basic Work Zone Staging Scenarios in the manual provide the Designer with the ability to quickly evaluate minimum width requirements for various lane closure scenarios.

High impact projects are those projects that have traffic demands that exceed work zone capacities for a period of more than six hours per 24 hour day. Moderate impact projects are those projects that have traffic demands that exceed work zone capacities for a period of zero to six hours per 24 hour day. Low impact projects are those projects that have traffic demands that do not exceed work zone capacities during any time of the 24 hour day.

Table. Preliminary Traffic Mitigation Impact Worksheet

Project: ______________________	**Overall Impact Assessment** ______________________

Project Characteristic	**Project Data (Qualitative)**	**Project Impact Level**
Primary Project Characteristics		
Road user cost		
Retail business disruption		
Neighborhood disruption		
Disruption to employers/ activity centers		
Political sensitivity		
Impacts of local/ developer projects		
Secondary Project Characteristics		
Duration of contract		
No. of construction stages		
Capacity reduction (%)		
Roadway type		
ADT		
Seasonal traffic increase		
Expected delay		
Project length		
Urbanization		
Truck volume		
Public/media attention		

The Department recognizes the public does not like to experience traffic delays due to construction. Therefore, the Department should include an alternative that introduces no traffic queues (*i.e.* volumes are less than capacity) in its evaluation matrix. Common methods to avoid queues are modifying the allowable lane closure hours and utilizing shoulder areas as temporary lanes. In the absence of accurate traffic data or in addition to the quantitative impact analysis, traffic impacts can also be assessed qualitatively.

A worksheet that may be used as a tool to identify impact level for a particular project. It provides space to note details for the primary project characteristics and to note secondary characteristics that might be important in selecting TM strategies. The worksheet also includes a column to indicate the project level suggested by each characteristic individually and defines an overall project impact level drawn from the combination of the levels for various project characteristics. If many of the characteristics, or even a few of the primary characteristics, suggest the project has high impacts, the project likely should be classified at the higher level.

Not surprisingly, as the level of impact of the project increases, the level of TM needed and its likely cost also will increase.

Low Impact Projects

The TM needs for low impact projects generally can be adequately addressed by minimal and "standard" traffic controls used routinely in construction areas. These could include, for example:

- Fixed signs to alert motorists to the presence of construction activity ahead
- Traffic cones and temporary barriers to guide motorists through the construction area

Low Impact TM Costs

Low impact TM projects are generally assumed to have no additional costs beyond those incurred routinely for construction.

Moderate Impact Projects

For moderate impact projects, the standard TM strategies used in low impact projects should be applied and augmented. Additional strategies that might be appropriate for these projects could include:

- Enhanced traffic operations and control strategies, such as construction area screening, variable message signs, reduced travel speeds, temporary parking and turn restrictions, and pavement markings
- Modest public outreach targeted to residents, businesses, and community centers in the local area affected by the construction: for example, postings on web sites, and fax network for employers and others
- Detours without additional upgrades to alternate routes
- Construction activity restrictions/non-construction increases in capacity (*e.g.*, narrow lanes) to minimize peak period delays

- Development of incident management plan to reduce delays caused by accidents
- Modest TDM strategies such as rideshare marketing

Moderate Impact TM Costs

Moderate impact TM projects rely primarily on strategies, such as public outreach, traffic operations and control, and TDM, which do not require construction of temporary infrastructure. These strategies generally have modest cost but may add 15 per cent or more to the construction cost.

High Impact Projects

Finally, high impact projects will receive the most aggressive TM strategies. For longer duration projects, more costly strategies could be appropriate due to the high cumulative cost of delay and disruption. Thus, for these projects, some temporary capital improvements might be warranted to maintain capacity. Additionally, it becomes more important to deliver information to the public-at-large and to motorists traveling through the construction area, and to support the needs of businesses for employees and customer access. Therefore, in addition to the strategies noted above for lesser impact projects, critical impact projects might also warrant the following:

- Advanced traffic operations and control strategies, such as ramp metering, signal adjustments, reversible lanes, lane closure restrictions, and truck restrictions
- Extensive public outreach to the public-at-large and to targeted groups, including public information meetings, brochures, news releases, media advertising, appearances at community events, a public information center, and a telephone hotline
- Detours that require capital improvements to alternate routes
- Temporary capital improvements, such as crossovers, runarounds, temporary roadways, and temporary use of shoulders and median areas to maintain capacity
- Contracting strategies, such as A+B bidding, contractor incentive/ disincentive provisions, lane rental provisions, and project consolidation
- Expanded TDM strategies such as temporary transit service, free bus service, Park & Ride lots, TDM information, and assistance to employers

High Impact TM Costs

High impact TM projects generally implement the most aggressive levels of TM strategies. However, in many cases the construction costs for these high impact projects are also high. Therefore, the TM strategies, as a percentage of the construction cost, are typically the same or less than moderate impact projects.

TRAFFIC COUNTS AND TRAFFIC SURVEYS

TRAFFIC COUNT

Fig. Bike counter with display showing the number of bikes on the particular day and accumulative for the year, for one bike lane in Copenhagen.

A traffic count is a count of traffic along a particular road, either done electronically or by people counting by the side of the road. Traffic counts can be used by local councils to identify which routes are used most, and to either improve that road or provide an alternative if there is an excessive amount of traffic. Also, some geography fieldwork involves a traffic count. They are useful for comparing two or more roads, and also can be used alongside other methods to find out where the CBD of a settlement is located.

COUNTING METHODS

Fig. A traffic counter on BIA road J-9 in the United States

To permanently or temporarily monitor the usage of a road, an electronic traffic counter can be installed or placed to measure road usage continuously or for a short period of time. Most modern equipment called ATR's (Automatic Traffic Recorders) store count and/or classification data recorded in memory in a timestamp or interval fashion that can be downloaded and viewed in software or via a count display on some equipment. In some instances people either draw up a table and/or use a tally to keep a record of vehicles which pass manually as an alternative to ATR's.

Example Table from a Traffic Count

This is an example of a traffic count, showing the type of vehicle and the data collected at a particular place in each direction. As you can see, this data has been compiled into numbers from each direction and a total count. The original table (which was used out on the road) used data recorded in tally form.

Vehicle Type	Direction X	Direction Y	Total
Car	48	47	95
Truck	2	41	43
Van	5	7	12
Bus	1	0	1
Bike	3	1	4
Pedestrian	3	4	7

TRAFFIC COUNTER DEVICE

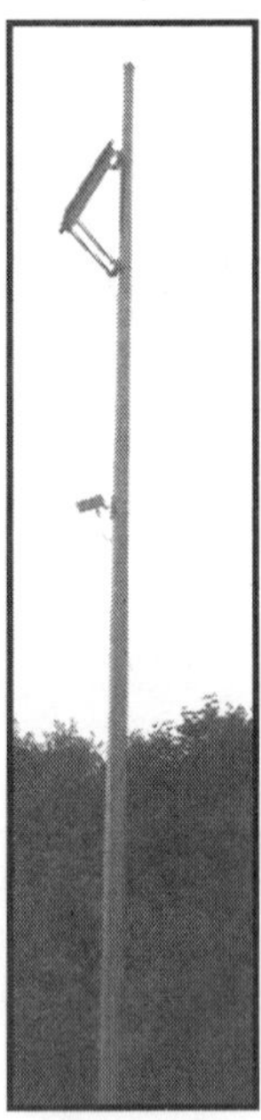

Fig. A radar-based traffic counter (about 2/3 of the way up the pole) powered by a solar panel (near top of pole).

A traffic counter is a device, often electronic in nature, used to count, classify, and/or, measure the speed of vehicular traffic passing along a given roadway. The device is usually deployed in near proximity to the roadway and uses an intrusive medium, such as pneumatic road tubes laid across the roadway, piezo-electric sensors embedded in the roadway, inductive loops cut into the roadway, or a combination of these to detect the passing vehicles. One of the first traffic counting units, called traffic recorders, was introduced in 1937 and operated off a strip laid across the street and operated off a six volt battery. Each hour it printed off a paper strip with the total for that hour. Recently, in the interest of worker safety and ease of installation, non-intrusive technologies have been developed. These devices generally use some sort of transmitted energy such as radar waves orinfrared beams to detect vehicles passing over the roadway.

Traffic counting methods

Transport planning at all levels requires understanding of actual conditions. This involves determination of vehicle or pedestrian numbers, vehicle types, vehicle speeds, vehicle weights, as well as more substantial information such as trip length and trip purpose and trip frequency. The first group of data dealing with the characteristics of vehicle or people movement is obtained by undertaking traffic counts. Those related to measuring trips involving knowledge of origin and destination require more detailed surveys.

There is a wide range of counting methods available. It is useful to distinguish between intrusiveand non-intrusive methods. The former include counting systems that involve placing sensors in or on the roadbed; the latter involve remote observational techniques. In general the intrusive methods are used most widely because of their relative ease of use and because they have been employed for decades. The only widely used non-intrusive method is manual counting, which enjoys wide application because of its ease. Intrusive methods, however, have evolved little over the last decade, but in the US, with federal transport policy emphasis on IT solutions to traffic management, progress is being made in the development of non-intrusive methods.The major intrusive methods include:

- *Bending plate:* A weight pad attached to a metal plate embedded in the road to measure axel weight and speed. It is an expensive device and requires alteration to the road bed.
- *Pneumatic road tube:* A rubber tube that is placed across the lanes that uses pressure changes to record the number of axle movements in a counter placed on the side of the road. The drawback is that it has limited lane coverage, may become displaced, and can be dislodged by snow ploughs.
- *Piezo-electric sensor:* A device that is placed in a groove cut into the roadbed of the lane(s) being counted. This electronic counter can be used to measure weight and speed. Cutting into the roadbed can affect the integrity of the roadbed and decrease the life of the pavement.

- *Inductive loop:* A wire embedded in the road in a square formation that creates a magnetic field that relays the information to a counting device at the side of the road. This has a generally short life expectancy because it can be damaged by heavy vehicles, and is also prone to installation errors.

The major non-intrusive methods include:

- *Manual observation:* A very traditional method involving placing observers at specific locations to record vehicle or pedestrian movements. At its simplest, observers use tally sheets to record, but numbers, on the other hand there are mechanical and electronic counting boards available that the observer can punch in each time an event is observed. It can record traffic numbers, type and directions of travel. Manual counts give rise to safety concerns, either from the traffic itself or the neighborhoods where the counts are being undertaken.
- *Passive and active infra-red:* A sensor detecting the presence, speed and type of vehicles by measuring infra-red energy radiating from the detection area. Typically the devices are mounted overhead on a bridge or pylon. The major limitation is the performance during inclement weather, and limited lane coverage.
- *Passive magnetic:* Magnetic sensors that count vehicle numbers, speed, and type are placed under or on top of the roadbed. In operating conditions the sensors have difficulty differentiating between closely spaced vehicles.
- *Microwave- Doppler/ Radar:* Mounted overhead the devices record moving vehicles and speed. With the exception of radar, devices they have difficulty in detecting closely spaced vehicles and do not detect stationary vehicles. They are not affected by weather.
- *Ultrasonic and passive acoustic:* Devices that sound waves or sound energy to detect vehicles. Those using ultrasound are placed overhead to record vehicle presence but can be affected by temperature and turbulence; the acoustic devices are placed alongside the road and can detect numbers and vehicle type.
- *Video image detection:* Use of overhead video cameras to record vehicle numbers, type and speed. Various software is available to analyze the video images. Weather may limit accuracy.

A recent study has examine the use of the various traffic count methods by State Departments of Transport in the US (Skszek 2001) found that less than half use any non-intrusive techniques. Part of the reason is the level of technical expertise required to operate the devices. Inductive loops are in use in all States, with very high levels of use (>90 per cent) for pneumatic rubber tubes and piezo-electronic road sensors. Manual counts were used by 82 per centof the States. In terms of satisfaction with the methods, manual counts and inductive loops were rated highest. Despite the poor acceptance of the non-intrusive devices, their cost effectiveness was shown to be higher than the inductive loops (Skszek 2001, 14). This suggests that the newer devices may gain wider use once their cost effectiveness becomes more widely appreciated.

Surveys

Traffic counts may provide some precise information about numbers of vehicles, their type, weight or speed, but they cannot provide other data that are essential in transport planning, such as trip purpose, routing, duration etc. Collecting these data requires more extensive survey instruments. These instruments include:

- *Mailed questionnaires:* Can include a wide range of questions; is relatively cheap to administer to large numbers of people, although preparation can be expensive; the main problem is the generally low response rate.
- *Travel diaries:* Soliciting respondents to keep a diary of the trips undertaken, times, purposes, modes etc.; extremely useful instrument constrained largely by the number of people willing to complete such a detailed inventory.
- *Telephone surveys:* With automated dialing this can achieve extensive coverage, but response rates are usually low.
- *Face-to-face home interviews:* Can overcome many of the errors based on misunderstanding of questions in mail surveys, but are extremely time-consuming and costly.

Extensive traffic surveys began to be developed in the 1950s. One of the earliest was the Chicago Area Transportation Study (CATS) that was undertaken in 1956, providing detailed O/D data on trip length, purposes, modes of travel, and travel patterns. This was followed in 1960 with the US Census's first attempt to collect journey to work (JTW) travel data in urban areas. Other metropolitan areas in the US and Canada, including Detroit and Toronto, copied and extended the scope of such surveys in the 1960s. The growth of surveys was encouraged by the results that provided the first comprehensive snapshots of urban travel activities in a society rapidly adopting the automobile and undertaking new types of travel behavior. This was a boon to transport planning. Furthermore, much of academic understanding of travel activity in cities has been drawn from these surveys. Since then national censuses in many countries have included travel surveys in their decennial inventories, and many planning agencies update and extend the results from the national surveys with local investigations.

All survey techniques represent a compromise between the objectives of the survey, the resources available, the coverage that is feasible, and the amount of data to be collected. The surveys instrument(s) that are employed depend largely on the resources available. Even national agencies find the costs of conducting national surveys are onerous. For example, it is estimated that the next daily trip survey to be undertaken in 2007 by the National Household Transport Survey (NHTS) in the US will be $14 million. Very common is the mail-back questionnaire.

CATS, for example, uses a questionnaire along with a travel diary, which involves sending out a letter of introduction to selected households, distribution of the questionnaire and instructions, mailing out reminder letters, and a telephone follow-up to selected individuals to verify their information. The NHTS survey of 2007 will be based on a national telephone survey.

The degree of detail required in most travel surveys means that even the largest agencies have to rely on sampling. It is usual to target households rather than individuals, since the household is a good predictor of travel behavior. Fixing the size of the sample is an extremely important issue. Sample size determines the degree of reliability of the results, but these have to be conditioned by the resources available and the survey instruments to be employed.

In its household surveys, CATS determined that 400 completed household responses would be sufficient to provide a statistically significant sample for each of the geographic units, and because it expected a 20 per cent rate of response, it could plan for the distribution of 2,000 questionnaires in each zone. A clustered random sample of approximately 2,000 addresses in each zone was taken. For national surveys in the US, samples of 26,000 households are sought. Because of national surveys may not provide a sufficiently reliable or detailed set of data for the needs of individual States or planning agencies, these agencies frequently 'back-on' additional counts in their areas when national surveys are undertaken.

CURRENT TRAFFIC DATA COLLECTION TECHNOLOGY

PRODUCT CLASSIFICATION

There are two main categories into which equipment for collecting traffic data can be placed—intrusive and non-intrusive devices. Intrusive (traditional) counting devices are those that involve placement of the sensor technology on top of or into the lane of traffic being monitored. They represent the most common devices used today including inductive loops, piezo-electric sensors and pneumatic rubber road tubes. Conversely, non-intrusive (non-traditional) counting devices such as passive acoustic and video image detection do not interfere with traffic flow either during installation or operation.

Within these two broad categories, thirteen different technologies were identified for classifying devices used for recording traffic data. The collection of count, speed, class, and weight-in-motion (WIM) data are the focus for this report.

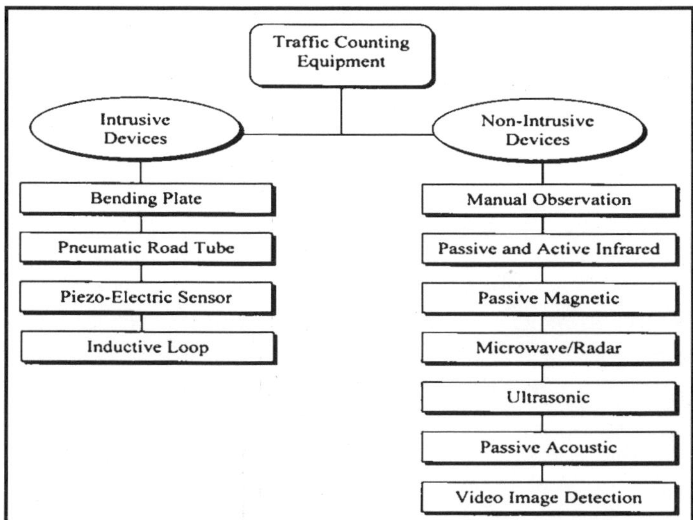

Fig. Product Classification

A definition of each category, as used for purposes of this report, is listed below

INTRUSIVE DEVICES

Bending Plate

Bending plate technology is most frequently used for collecting weight-in-motion data. The device typically consists of a weigh pad attached to a metal frame installed into the travel lane. A vehicle passes over the metal frame causing it to slightly "bend." Strain gauge weighing elements measure the strain on the metal plate induced by the vehicle passing over it. This yields a weight based on wheel/axle loads on each of two scales installed in a lane. The devices also is used to obtain classification and speed data.

Pneumatic Road Tube

A pneumatic road tube is a hollow rubber tube placed across the roadway that is used to detect vehicles by the change in air pressure generated when a vehicle tire passes over the tube. A device attached to the road tubes is placed at the roadside to record the change in pressure as a vehicle axle. Axle counts can be converted to count, speed, and/or classification depending on how the road tube configuration is structured.

Piezo-Electric Sensor

Piezo-electric sensors are mounted in a groove that is cut into the roadway surface within the traffic lane. The sensors gather data by converting mechanical energy into electrical energy. Mechanical deformation of the piezoelectric material causes a change in the surface charge density of the material so that a change in voltage appears between the electrodes. The amplitude and frequency of the signal is directly proportional to the degree of deformation. When the force of the vehicle axle is removed, the output voltage is of opposite polarity. The change in polarity results in an alternating output voltage. This change in voltage can be used to detect and record vehicle count and classification, weight-in-motion and speed.

Inductive Loop

An inductive loop is a wire embedded into or under the roadway in roughly a square configuration. The loop utilizes the principle that a magnetic field introduced near an electrical conductor causes an electrical current to be induced. In the case of traffic monitoring, a large metal vehicle acts as the magnetic field and the inductive loop as the electrical conductor. A device at the roadside records the signals generated.

NON-INTRUSIVE DEVICES

Manual Observation

Manual observation involves detection of vehicles with the human eye and hand recording count and/or classification information. Hand-held devices are available for on-site recording of information gathered by one or more individuals observing traffic.

Passive and Active Infrared

Passive infrared devices detect the presence of vehicles by measuring the infrared energy radiating from the detection zone. A vehicle will always have a temperature contrast to the background environment. The infrared energy naturally emanating from the road surface is compared to the energy radiating when a vehicle is present. Since the roadway may generate either more or less radiation than a vehicle, the contrast in heat energy is detected. The possibility of interference with other devices is minimized because the technology is completely passive. Passive infrared detectors are typically mounted directly over the lane of traffic on a gantry, overpass or bridge or alternatively on a pole at the roadside.

Active infrared devices emit a laser beam at the road surface and measure the time for the reflected signal to return to the device. When a vehicle moves into the path of the laser beam the time it takes for the signal to return is reduced. The reduction in time indicates the presence of a vehicle. The mounting position for active infrared detectors is more variable. The Autosense devices from Schwartz Electro-Optics, Inc. are mounted over the lane(s) of traffic to be monitored or in a side-fire mount perpendicular to the lane of traffic. There also are portable, devices that are placed roadside so the laser beams are directed parallel to the road surface across the lane of traffic. Both active and passive infrared devices can be used to record count, speed, and classification data.

Passive Magnetic

Passive magnetic devices detect the disruption in the earth's natural magnetic field caused by the movement of a vehicle through the detection area. In order to detect this change the device must be relatively close to the vehicles. This limits most applications to installation under or on top of the pavement, although some testing has been done with side fire devices in locations where they can be mounted within a few feet of the roadway. Magnetic sensors can be used to collect count, speed, and classification data.

Microwave - Doppler/Radar

Doppler microwave detection devices transmit a continuous signal of low-energy microwave radiation at a target area on the pavement and then analyze the signal reflected back. The detector registers a change in the frequency of waves occurring when the microwave source and the vehicle are in motion relative to one another. According to the Doppler principle, when a moving object reflects the radar beam emitted from the detector, the frequency of the reflected wave is changed proportionally to the speed of the reflecting object. This allows the device to detect moving vehicles and determine their speed. The only sensors identified using Doppler microwave are produced by Microwave Sensors, Inc. and are used primarily as a detection device designed to trigger operation of a traffic controller. In this capacity, they are placed in an overhead mounting position.

Radar (radio detecting and ranging) is capable of detecting distant objects and determining their position and speed of movement. With vehicle detection, a device directs high frequency radio waves, either a pulsed, frequency-modulated or phase-modulated signal, at the roadway to determine the time delay of the return signal, thereby calculating the distance to the detected vehicle. Radar devices are capable of sensing the presence of stationary vehicles. They are insensitive to weather and provide day and night operation. The device is placed in a side-fire mount off the shoulder of the roadway. This technology is capable of recording count, speed, and classification data.

Ultrasonic and Passive Acoustic

Ultrasonic devices emit pulses of ultrasonic sound energy and measure the time for the signal to return to the device. The sound energy hits a passing vehicle and is reflected back to the detection device. The return of the sound energy in less time than the normal road surface background is used to indicate the presence of a vehicle. Ultrasonic sensors are generally placed over the lane of traffic to be monitored. Passive acoustic devices utilize sound waves in a somewhat different manner. These systems consist of a series of microphones aimed at the traffic stream. The device detects the sound from a vehicle passing through the detection zone. It then compares the sound to a set of sonic signatures preprogrammed to identify various classes of vehicles. The primary source of sound is the noise generated by the contact between the tire and road surface. These devices are best used in a side fire position, pointed at the tire track in a lane of traffic to collect count, speed, and classification data.

Video Image Detection

Video image detection devices use a microprocessor to analyze the video image input from a camera. Two techniques, trip line and tracking, are used to record traffic data. Trip line techniques monitor specific zones on the roadway to detect the presence of a vehicle. Video tracking techniques employ algorithms to identify and track vehicles as they pass through the field of view. Different manufacturers technology may employ one or both of these techniques. Optimal mounting position for video image detectors is directly over the lane(s) to be monitored with an unobstructed view of traffic. Side mounting is feasible but large vehicles may obstruct detection zones. The mounting height is related to the desired lane coverage, usually 35 to 60 feet above the roadway. Video detection devices are capable of recording count, speed, and classification data.

MANUFACTURERS OF TRAFFIC COUNTING DEVICES

A list of manufacturers was compiled through an Internet search and by conversations with traffic counting professionals. Any manufacturer producing one or more devices for collection of count, speed, classification, and/or WIM data was considered. Many systems are "open systems" in that the sensors and data collection devices may be from different manufacturers. This is the case with most pneumatic rubber tube and inductive loop systems. In addition, the data collection devices employed by these systems may utilize more than one sensor type. For the most part, the more sophisticated the technology, the more likely the system will be a "closed" system.

Table contains manufacturers identified by this researcher who were cooperative in supplying detailed product information and were responsive to questions regarding their products. The devices are categorized by their sensor technology; however, the product listing is limited to the devices used to interpret data output from the sensors. Manufacturers producing only sensors and not data recording devices were excluded from Table. A more detailed listing that includes contact name, address, telephone number, e-mail address, and website information for each manufacturer is included as Appendix E.

- 3M, Intelligent Transportation Systems
  - ASIM Technologies, Ltd.
  - ATD Northwest
  - Boschung America
  - Computer Recognition Systems, Inc.
  - Diamond Traffic Products
  - Econolite Control Products, Inc.
  - EFKON AG
  - Electronic Integrated Systems, Inc.
  - Electronique Controle Mesure (ECM)
  - Eltec Instruments, Inc.
  - Golden River TRAFFIC, Ltd.
  - International Road Dynamics Inc.
  - International Traffic Corp./ Pat America
  - Iteris (formerly Odetics)
  - JAMAR Technologies, Inc.
  - Measurement Specialties, Inc.
  - MetroCount
  - Mikros Systems (Pty.), Ltd.
  - Mitron Systems Corporation
  - Nestor Traffic Systems, Inc.
  - Nu-Metrics
  - Peek Traffic Inc. - Sarasota
  - Reno Detection Systems
  - Schwartz Electro-Optics, Inc.
  - SmarTek Systems, Inc.
  - Spectra-Research
  - Traficon
  - U.S. Traffic Corporation

PRODUCT INFORMATION

Each manufacturer was contacted for product information for any device they distribute that is used to collect traffic count, speed, classification, and/ or WIM data. The focus was on devices designed specifically for use in high speed, freeway applications. Devices used primarily for presence detection at intersections for traffic signal applications or on freeway entrance ramps for traffic management were not considered.

Table summarizes devices currently on the market including manufacturer name, sensor type, and data collected. Although the devices are listed by sensor type, the emphasis was on acquiring information about the data recording and interpretation equipment that is attached to the various sensors. The sensor type used with a particular piece of equipment may or may not be made by the manufacturer listed. As previously stated, there is a wide range of open and closed systems available. Devices used for recording information obtained by manual observation were not included.

AUTOMATIC COUNT METHOD

The automatic count method provides a means for gathering large amounts of traffic data. Automatic counts are usually taken in 1-hour interval for each 24-hour period. The counts extend for a week, month, or year. When the counts are recorded for each 24-hour time period, the peak flow period can be identified. Automatic counts are recorded using one of three methods: portable counters, permanent counters, and videotape.

Vehicle Classification

Traffic volumes vary over time on all roads. Traffic volumes also vary dramatically from one road to another. These variations in traffic volume are even more apparent when volumes for specific vehicle types (classification) are analyzed. Consequently, the vehicle classification data collection program must gather sufficient data on traffic patterns of important vehicle types to accurately quantify the truck traffic stream to meet the needs of users. These include; time of day, day of week, time of year, direction.

Vehicle classification counts are used in establishing structural and geometric design criteria, computing expected highway user revenue, and computing capacity. If a high percentage of heavy trucks exist or if the vehicle mix at the crash site is suspected as contributing to the crash problem, then classification counts should be conducted. Typically cars, station wagons, pickup and panel trucks, and motorcycles are classified as passenger cars. The observer records the classification of vehicles and its direction of travel at the intersection.

Integration of Classification Count

The vehicle classification counts required should not be considered separate from the volume counts traditionally performed. Instead, they should be integrated with the traditional volume counts. Because classification counts provide both classification and total volume information, they can replace traditional volume counts reducing duplication and error. Traffic surveillance equipment is used as part of advanced traffic management systems (ATMS) or advanced traveller information systems (ATIS) can be used to supply both total volume and vehicle classification information. Intelligent transportation system (ITS) technology and its resulting data are often present at high profile locations as part of safety enhancement systems. These systems can supply useful, continuous traffic monitoring data. Coordinating these traffic monitoring activities can lead to significant improvements in the amount of data available to users, while at the same time reducing the cost of data collection.

Uses of Classification Data

Vehicle classification data are of considerable use to agencies involved in almost all aspects of transportation planning and engineering. The need for information on truck volumes and freight movements is growing with the recognition of role that freight mobility plays in the economy, and as highway engineers realize the importance of truck volume and operating characteristics on the geometric and structural design of roadways and bridges.

- Pavement design
- Pavement management
- Scheduling the resurfacing, reconditioning, and reconstruction of highways
- Prediction and planning for commodity flows and freight movements
- Development of weight enforcement strategies
- Vehicle crash record analysis
- Environmental impact analysis
  Analysis of alternative highway regulatory and investment policies.

Vehicle Occupancy

Vehicle occupancy measurement is an important part of transportation congestion management and it is used for evaluating the efficiency of road system, High Occupancy Vehicle (HOV) lanes or particular congestion reduction programs. The measure occupancy is a function of speed and length of individual vehicle and thus, it could consider the effects of varying vehicle length and speed. Hence, it can be considered as a logical substitute of density. In other words, occupancy, based on practical consideration, is defined as the percentage of time the detection zone is occupied by the vehicles. Therefore, occupancy measured using detectors depends on the length of detection zone, each detector type has a differing zone of influence (detector length) and the zone of influence is effectively added to vehicle length. Hence, the measured occupancy may be different for different detection zones even for the same site having identical traffic, depending on the size and nature of the detectors. Development of intelligent systems that extract traffic density and vehicle classification information from traffic surveillance systems is crucial in traffic management. It is important to know the traffic density of the roads real time especially in HOV lanes for effective traffic management. Time estimation of reaching from one location to another and recommendation of different route alternatives using real time traffic density information are very valuable for metropolitan city residents.

Travel Time

Travel time can be defined as the period of time to transverse a route between any two points of interest. It is a fundamental measure in transportation. Travel time is also one of the most readily understood and communicated measure indices used by a wide variety of users, including transportation engineers, planners, and consumers. Travel time data is useful for a wide range of transportation analyses including congestion management, transportation planning, and traveller information. Congestion management systems commonly use travel time-based performance measures to evaluate and monitor traffic congestion. In addition, some metropolitan areas provide real-time travel time prediction as part of their advanced traveller information systems (ATIS). Travel time data can be obtained through a number of methods. Some of the methods involve direct measures of travel times along with test vehicles, license plate matching technique, and ITS probe vehicles. Additionally, various sensors (*e.g.* inductance loop detectors, acoustic sensors) in ITS deployment collect a large amount of traffic data every day, especially in metropolitan areas. Such data can be used for travel time estimation for extensive applications when direct measurements of travel times are not available.

Delay

The delay defines as "The additional travel time experienced by a driver, passenger, or pedestrian". Delay is thus the difference between an "ideal" travel time and "actual" travel time. Since the definition of delay depends on a hypothetical "ideal travel time", delay is not always directly measurable in the field. If the ideal travel time is defined as off-peak travel time, then the measured delay is difference between the actual measured travel time during peak period, and the actual measured travel time during off-peak period. If the ideal travel time is defined as travel at the posted speed limit, then the delay cannot be directly measured in the field. It is estimated by subtracting the hypothetical travel time at the posted speed limit from the measured mean travel time in the field.

DETECTOR TECHNOLOGY

General

In traffic detector information is derived from technologies divided into two main groups, information collected via in-situ detectors, deployed at location of interests, or information from mobile technologies that are located within vehicles themselves. Over the last two decades, there has been an increase in the provision of services that are specific to vehicle types, as well as fleet or asset management and tracking, based on in-vehicle technologies. In-vehicle technologies have really come into realization through the advantage of satellite-based technologies, and are perceived as playing an increasing role in the future.Such technologies not only improve our ability to manage networks efficiently, but will also have a direct impact of the types of policy instruments available to authorities, the operation of so-called ITS.

In-Situ Technologies

In-situ traffic detector technologies are further divided into two categories: *Intrusive* technologies that are physically mounted at, or below, the road surface, installation of which causes potential disruption to traffic. Conversely, *non-intrusive* technologies are mounted at, or above the road surface, and their installation causes little or no disruption to traffic. Detectors of both types temporary or permanent nature, though sub-surface *intrusive* installations are, by necessity, usually permanent. All in-situ detectors will provide some measure of the volume of vehicle flow. Particular detector technologies will vary as to their reliability of the flow estimate, and their ability to provide accurate additional information on vehicle category or speed. A single sensor gives only flow or occupancy information. Two adjacent sensors are required for speed or classification assessment. The time-lag and separation distance between the onset of consecutive events at the sensors have been used to estimate vehicle speed. Classification information is derived either from vehicle length or through examination of the form of the profile generated as output from the sensor.

INTRUSIVE TECHNOLOGIES

Typical examples of intrusive technologies, their sensor types and installation locations are shown in Fig. 1. The first types of units are passive magnetic or magnetometer sensors that are either permanently mounted within holes in the road, or affixed to the road surface in some fashion. The unit communicates to a nearby base station processing unit using either wires buried in the road, or wireless communications. The sensor has a circular or elliptically offset *zone of detection* (*i.e.*, the blue area).

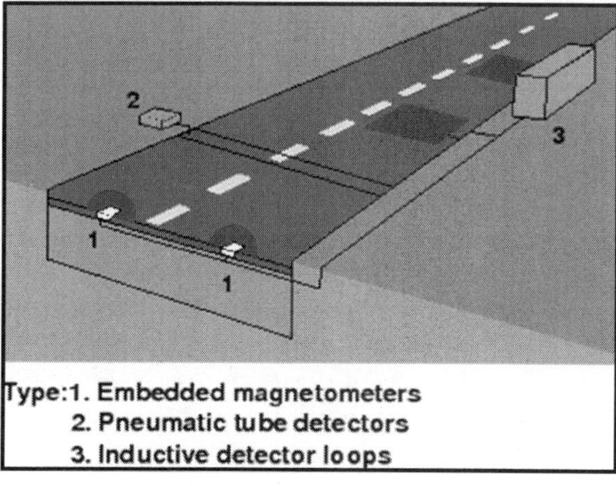

Fig. Typical intrusive detector configurations.

The second types of units use pneumatic tubes that are stretched across the carriageway and affixed at the kerb side at both ends. Such systems are only be deployed on a temporary basis, due to the fragile nature of tubes, which are easily damaged or torn up by heavy or fast moving vehicles. The third type are inductive detector loops (IDL), consisting of coated wire coils buried in grooves cut in the road surface, sealed over with bituminous filler. A cable buried with the loop sends data to a roadside processing unit. The zone of detection for inductive loop sensors depends on the cut shape of the loop slots.

The zones depending on the overall sensitivity of system not correspond precisely to the slot dimensions. IDLs are a cheap and mature technology. They are installed on both major roads and within urban areas, forming the backbone detector network for most traffic control systems. The fourth type of intrusive system is Weigh-In-Motion (WIM), detectors that consist of a piezoelectric sensor (*e.g.* 'bending-plate' or fiber-optic) system laid in a channel across the road.

These systems are relatively rare and are used in specific locations for enforcement or access control. They are usually coupled with other systems, either intrusive or non-intrusive, to provide additional cross-checks on collected data.

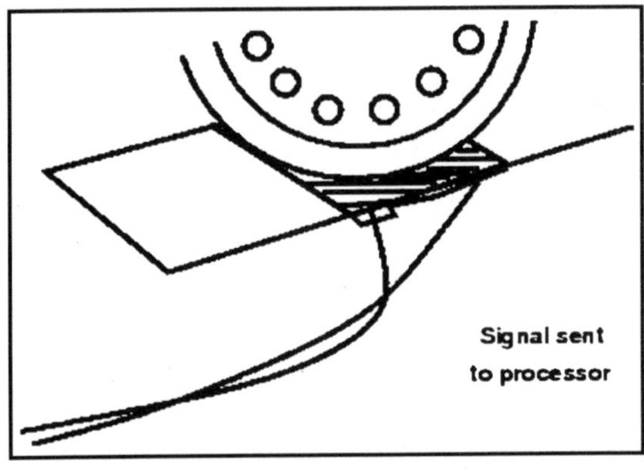

Fig. Weigh-In-Motion Detector system, Source

Pneumatic Tube Detector

Pneumatic road tube sensors send a burst of air pressure along a rubber tube when a vehicles tire passes over the tube. The pulse of air pressure closes an air switch, producing an electrical signal that is transmitted to a counter or analysis software. The pneumatic road tube sensor is portable, using lead-acid, gel, or other rechargeable batteries as a power source. The road tube is installed perpendicular to the traffic flow direction and is commonly used for short-term traffic counting, vehicle classification by axle count and spacing. Some data to calculate vehicle gaps, intersection stop delay, stop sign delay, and saturation flow rate, spot speed as a function of vehicle class, and travel time when the counter is utilized in conjunction with a vehicle transmission sensor.

Advantages

- Cheap and self-contained, the easiest to deploy of all intrusive systems, recognized technology with acceptable accuracy for strategic traffic modeling purposes, hence very widely used.
- Axle-based classification appears attractive, given sub-vehicle categories are partially axle based.

Disadvantages

- Some units are not counted or classify vehicles.
- Tube installations are not durable, the life of tubes are less than one month only.
- The tube detectors are not suitable for high flow and high speed roads.
- Units should not be positioned where there is the possibility of vehicles parking on the tube.
- It can't detect the two wheelers.

Inductive Detector Loop (IDL)

Oscillating electrical signal is applied to the loop. The metal content of a moving vehicle chassis changes the electrical properties of circuit. Changes are detected at a roadside unit, triggering a vehicle event. A single loop system collects flow and occupancy. The speed can be calculated by the assumptions that are made for the mean length of vehicles. Two-loop systems collect flow, occupancy, vehicle length, and speed.

Advantages

- It is a very cheap technology. Almost every dynamic traffic control system in this world uses IDL data.

Disadvantages

- Loops are damaged by utility and street maintenance activities or penetration of water.
- IDLs with low sensitivity fail to detect vehicles with speed below a certain threshold, and miscount vehicles with complex or unusual chassis configurations, or vehicles with relatively low metal content (*e.g.* motorcycles).
- IDL data supplied to traffic control systems have a very low sample rate.
- Not suitable for mounting on metallic bridge decks.
- Some radio interference occurs between loops in close proximity with each other.

Magnetometers/Passive magnetic systems

Magnetometers monitor for fluctuations in the relative strength of the

Earth's magnetic field, which is changed by the presence of a moving metal object *i.e.,* a vehicle. A single passive magnetic system collects flow and occupancy. Two magnetometer systems collect flow, occupancy, vehicle length, and speed. Two types of magnetic field sensors are used for traffic flow parameter measurement. The first type, the two-axis fluxgate magnetometer, detects changes in vertical and horizontal components of the Earth s magnetic field produced by a ferrous metal vehicle. The two-axis fluxgate magnetometer contains a primary winding and two secondary sense windings on a coil surrounding high permeability soft magnetic material core. The second type of magnetic field sensor is the magnetic detector, more properly referred to as an induction or search coil magnetometer shown in Fig. 3. It detects the vehicle signature by measuring the change in the magnetic lines of flux caused by the change in field values produced by a moving ferrous metal vehicle. These devices contain a single coil winding around a permeable magnetic material rod core. However, most magnetic detectors cannot detect stopped vehicles, since they require a vehicle to be moving or otherwise changing its signature characteristics with respect to time.

Advantages

- More usually mounted in a small hole in road surface and hardwired to the processing unit. Suitable for deployment on bridges.

Disadvantages

- Possibly damaged by utility maintenance activities, as with IDLs.

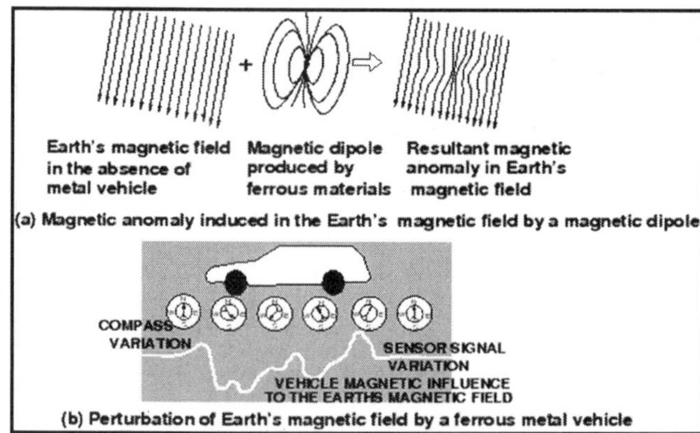

Fig. Weigh-In-Motion Detector system.

Weigh-In-Motion (WIM) systems

Bending Plate Bending plate WIM systems utilize plates with strain gauges bonded to the underside. The system records the strain measured by strain gauges and calculates the dynamic load. Static load is estimated using the measured dynamic load and calibration parameters. Calibration parameters account for factors, such as vehicle speed and pavement or suspension dynamics that influence estimates of the static weight. The accuracy of bending plate WIM systems can be expressed as a function of the vehicle speed traversed over the plates, assuming the system is installed in a sound road structure and subject to normal traffic conditions. *Advantages* Bending plate WIM systems is used for traffic data collection as well as for weight enforcement purposes. The accuracy of these systems is higher than piezoelectric systems and their cost is lower than load cell systems. Bending plate WIM systems do not require complete replacement of the sensor. *Disadvantages* Bending plate WIM systems are not as accurate as load cell systems and are considerably more expensive than piezoelectric systems.

Piezoelectric Piezoelectric WIM systems contain one or more piezoelectric sensors that detect a change in voltage caused by pressure exerted on the sensor by an axle and thereby measure the axle s weight. As a vehicle passes over the piezoelectric sensor, the system records the sensor output voltage and calculates the dynamic load. With bending plate systems, the dynamic load provides an estimate of static load when the WIM system is properly calibrated. The typical piezoelectric WIM system consists of at least one piezoelectric sensor and two ILDs. The piezoelectric sensor is placed in the travel lane perpendicular to the travel direction. The inductive loops are placed upstream and downstream of the piezoelectric sensor. The upstream loop detects vehicles and alerts the system to an approaching vehicle. The downstream loop provides data to determine vehicle speed and axle spacing based on the time it takes the vehicle to traverse the distance between the loops. Fig. 4 shows a full-lane width piezoelectric WIM system installation. In this example, two piezoelectric sensors are utilized on either side of the downstream loop. *Advantages* Typical piezoelectric WIM systems are among the least expensive systems in use today in terms of initial capital costs and life cycle maintenance costs. Piezoelectric WIM systems can be used at higher speed ranges (16 to 112 kmph) than other WIM systems. Piezoelectric WIM systems can be used to monitor up to four lanes. *Disadvantages* Typical piezoelectric systems are less accurate than load cell and bending plate WIM systems. Piezoelectric sensors for WIM systems must be replaced at least once every 3 years.

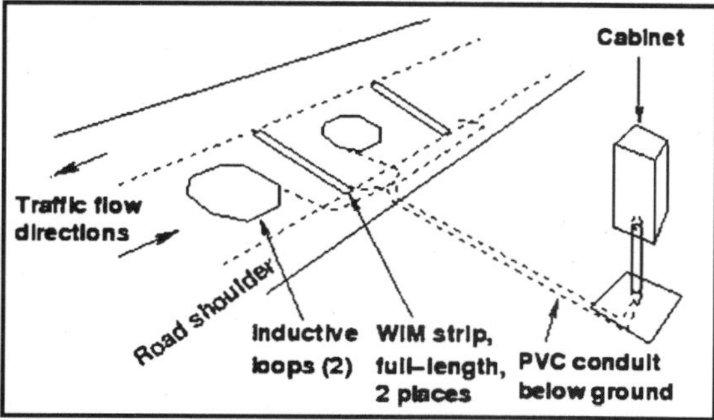

Fig. WIM installation with full-length piezoelectric sensors.

NON-INTRUSIVE TECHNOLOGIES

Non-intrusive technologies include video data collection, passive or active infrared detectors, microwave radar detectors, ultrasonic detectors, passive acoustic detectors, laser detectors and aerial photography. All these technologies represent emergent fields that are expanding rapidly with continuing advances in signal processing. At present time such technologies are used to provide supplemental information for selected locations or for specific applications (*e.g.*, queue detection at traffic signals). Most non-intrusive systems are operationally and somewhat visually similar, consisting of small electronics unit mounted in a weatherproof housing placed in various locations. The first type of non-invasive detectors are roadside mast-mounted. The detector possesses a field-of-regard covering an oblique area upstream or downstream of the unit. There are also multiple zones of detection defined within the overall field of regard, or the overall zone of detection same as the field of regard, depending on the specific detector type and technology. Obscuration problems occur when high-sided vehicles screens lower vehicles from the detector or the field-of-view being too large, leading to detection of vehicles outside the desired lane.

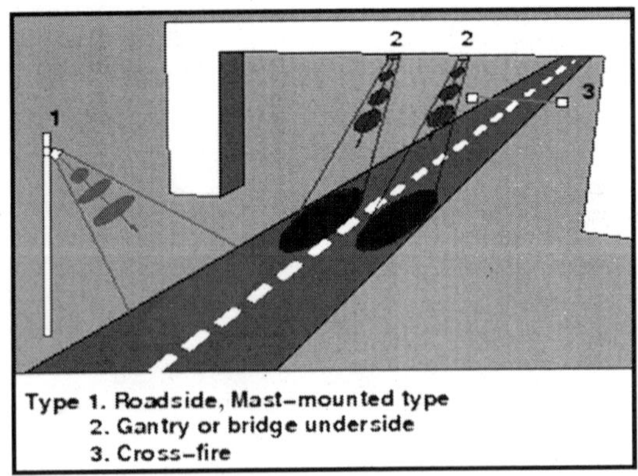

Fig. Typical non-intrusive technology configurations

The second type of non-invasive detectors are mounted on gantries or bridge undersides, with field of regard directly below, or at a slight oblique to the unit. Finally, some units, such as open-path pollutant monitors are mounted road side at ground level, firing a beam across the road. Such units are subject to side-by-side masking and hence most suitable for only single lane, unidirectional flows.

Video image detection (VID)

The traffic parameters are collected by frame-by-frame analysis of video images captured by roadside cameras. The following parameters are collected: Depending on the processing methodology almost all traffic parameters are captured from video analysis. Simple video systems often collect flow volume and occupancy. More complex systems allow the extraction of further parameters.

Advantages: Possibility to capture all desired traffic information, including some parameters that are not readily obtainable using other types of detectors Possibility of a permanent visual record of the traffic flow that reviewed and analyzed by a human operator.

Disadvantages: VID systems are susceptible to obscure issues, as with other non-intrusive detectors. Performance of VID systems might be degraded in bad weather or low light conditions.

Infrared Sensors

The sensors are mounted overhead to view approaching or departing traffic or traffic from a side-looking configuration. Infrared sensors are used for signal control; volume, speed, and class measurement, as well as detecting pedestrians in crosswalks. With infrared sensors, the word detector takes on another meaning, namely the light-sensitive element that converts the reflected or emitted energy into electrical signals. Real-time signal processing is used to analyze the received signals for the presence of a vehicle.

- Passive Infrared (PIR) Detection of vehicle based on emission or reflection of infrared (electromagnetic radiation of frequency 10^{11} — 10^{14} H_z) radiation from vehicle surface, as compared to ambient levels emitted or reflected from the road surface shown in Fig. 6. The PIR system collected following parameters: Flow volume, Vehicle presence, and detection zone occupancy. Speed with unit with multiple detection zones.

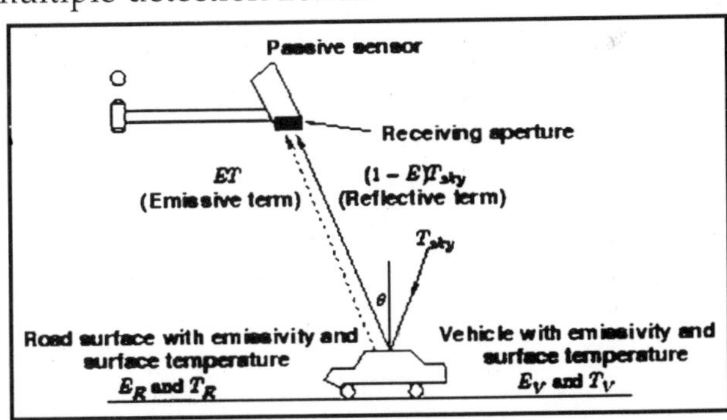

Fig. Emission and reflection of energy by vehicle and road surface.

Advantages

- Relatively long wavelength of light used in PIR systems makes them less susceptible to weather effects.

Disadvantages

- Accuracy of speed information is poor with low resolution sensors. Vehicle length determination is highly problematic for the same reason.

Active Infrared (AIR)/Laser Low power LED or laser diode fires a pulsed or continuous beam down to road surface as shown in Fig. 7. Time for reflection to return is measured. Presence of a vehicle lowers the time of reflection. High scanning rates provides a detailed profile for classification determination. Use of Doppler frequency shift from moving object allows for very accurate speed determination. The AIR system collected following parameters flow volume, speed, classification, vehicle presence, traffic density.

Advantages

- Very accurate flow, speed and classifications possible.
- Laser systems work in day and night conditions.

Disadvantages

- Active near-IR sensors adversely affected by weather conditions.
- Laser systems impeded by haze or smoke.
- Some problems with tracking small vehicles reported.
- Relatively high costs compared to other units. Precise, but limited zone of detection require additional units over other systems.

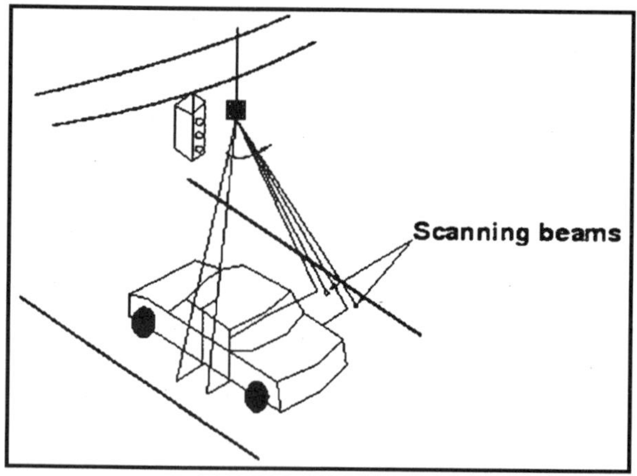

Fig. Laser radar beam geometry.

Microwave - Doppler and Radar

Low energy microwave radiation (2.5 to 24 GHz) is transmitted into the detection zone. Objects within the zone reflect a portion of the radiation back to a receiver. Doppler units use the frequency shift of the return to calculate speed. It can't detect the stationary objects. The microwave system collected following parameters. Doppler - Flow volume and speed; Frequency-Modulated, Continuous Wave (FMCW) - Flow volume, speed and presence; Microwave - Flow volume, speed, presence, possibly classification;

Advantages

- Very accurate. Easy to install, long ranged.
- Multiple detection zones possible.
- Day or night operation.

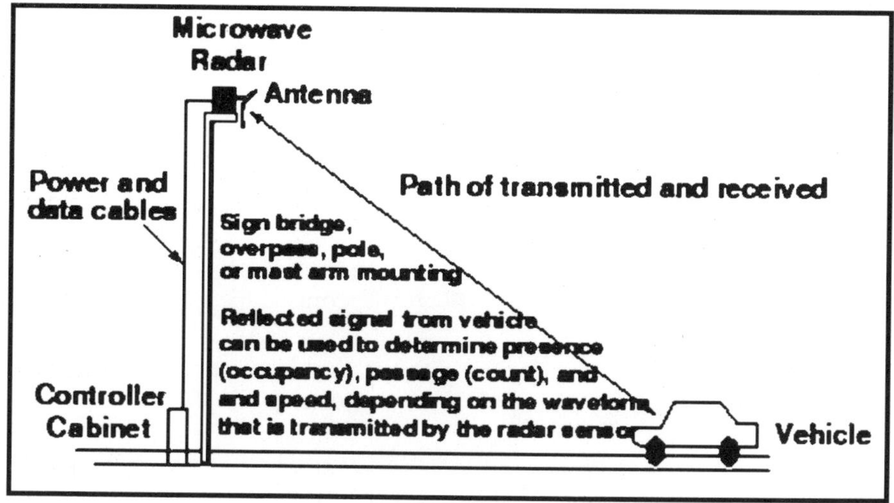

Fig. Microwave radar operation.

Disadvantages

- Possible sensitivity to spurious returns from adjacent objects
- Restrictions on use due to electromagnetic interference with other electronics.

Pulsed and Active Ultrasonic

Ultrasonic sensors transmit pressure waves of sound energy at a frequency between 25 and 50 KHz. Pulse waveforms measure distances to the road surface and vehicle surface by detecting the portion of the transmitted energy that is reflected towards the sensor from an area defined by the transmitter's beam width.

When a distance other than that to the background road surface is measured, the sensor interprets that measurement as the presence of a vehicle as shown in Fig.

The received ultrasonic energy is converted into electrical energy that is analyzed by signal processing electronics that is either collocated with the transducer or placed in a roadside controller. Vehicles flow and vehicular speed can be calculated by recording the time at which the vehicle crosses each beam.

Advantages

- Highly accurate.

Disadvantages

- Environmental effects affecting sound propagation degrade performance.
- Pulsed units with low sampling rate miscount or misclassify fast moving vehicles.

Fig. Ultrasonic range-measuring sensors.

Passive Acoustic Array Sensors

An array of microphones is used to detect the sound of an approaching vehicle above an ambient threshold level. Time lags and signal variations between microphone positions are used to determine vehicle location relative to the array. Further processing of signal yield to speed information and possibly engine type classification. It collected flow, speed, occupancy, possibly classification.

Advantages

- Completely passive system
- Direct speed measurement.

Disadvantages

- Environmental effects affecting sound propagation degrade performance
- Low accuracy in busy locations due to interference from adjacent sources.

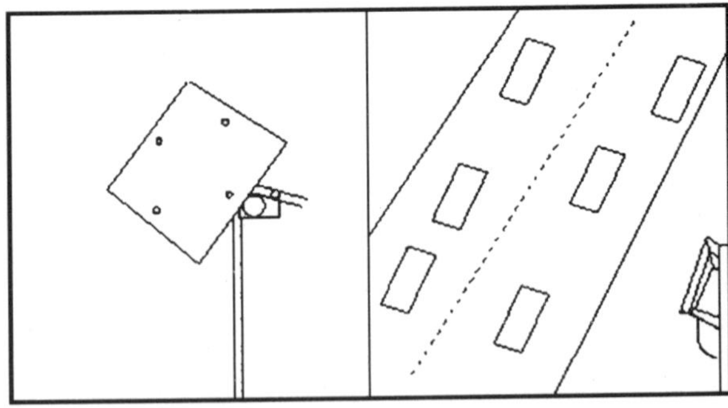

Fig. Acoustic array sensors.

IN-VEHICLE TECHNOLOGIES OR FLOATING CAR DATA (FCD)

In addition to using in-situ technologies, many network management applications make use of in-vehicle devices, generically termed Automatic Vehicle Location (AVL) systems. AVL devices either provide positional information whenever a suitably equipped vehicle passes a certain point in the network, or continuous information as the vehicle travels through a network. The former system typically relies on appropriate vehicles being equipped with transponders which transmit and receive information from roadside units. The latter system uses vehicles equipped with Global Positioning System (GPS) technology.

The principle of FCD is to collect real-time traffic data by locating the vehicle via mobile phones or GPS over the entire road network as shown in Fig. 11. It represents that all vehicles are equipped with mobile phone or GPS which will act as a sensor for the road network. Data such as car location, speed and direction of travel are sent anonymously to a central processing centre. After collecting and extracting, useful information such as status of traffic and alternative routes it can be redistributed to the drivers on the road. FCD is an alternative or rather complement source of high quality data to existing technologies. They will help improve safety, efficiency and reliability of the transportation system. They are becoming crucial in the development of ITS.

GPS-based FCD

GPS is becoming more and more useful and inexpensive; few cars had been equipped with GPS system and were made to pass a certain point in the network. The vehicle location precision was found to be relatively high, typically less than 30m. Generally, traffic data obtained from private vehicles or trucks are more suitable for motorways and rural areas.

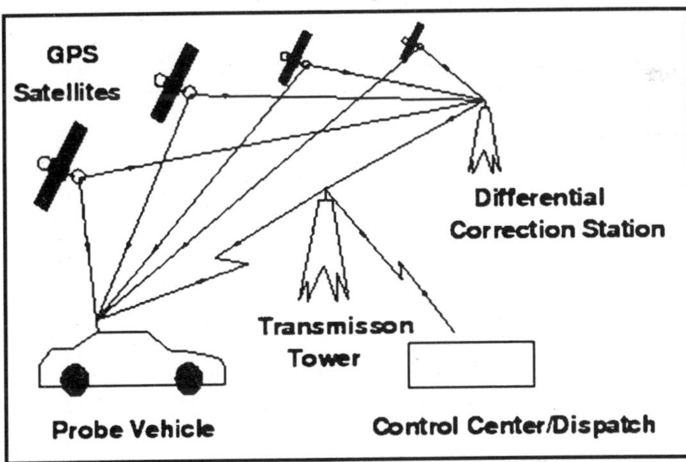

Fig. Communication from GPS, source

Currently, GPS probe data are widely used as a source of real-time information by many service providers but it suffers from a limited number of vehicles equipped and high equipment costs compared to floating cellular data.

Radio-frequency identification (RFID) or Transponder Systems

Radio-frequency identification (RFID) is an automatic identification method, relying on storing and retrieving data from remote areas using devices called RFID tags or transponders. The technology requires some extent of cooperation of an RFID reader and an RFID tag. An RFID tag is an object that can be applied to or incorporated into a product, animal, or person for the purpose of identification and tracking using radio waves. Some tags can be read from several meters away and beyond the line of sight of the reader. A basic RFID system consists of three components:

1. An antenna or coil
2. A transceiver (with decoder)
3. A transponder (RF tag)

An RFID tag is comprised of a microchip to collect information and an antenna that transmits this data wirelessly to a reader. At its most basic, the chip will contain a serialized identifier, or license plate number, that uniquely identifies that item. Typically, processed data would be used to provide revised scheduling and arrival time information to the general public, via variable information signs. Transponder systems are also used with Selective Vehicle Detection (SVD) systems which are designed to allow priority at traffic signals or cordon points for public transport or emergency service vehicles.

Typical Applications for RFID

- Automatic Vehicle identification
- Inventory Management
- Work-in-Process
- Container or Yard Management
- Parking Management

Advantages

- RFID tags can be read through materials without line of sight.
- RFID tags can be read automatically when a tagged product comes past or near a reader.

Disadvantages

- Reader collision occurs when the signals from two or more readers overlap.
- The tag is unable to respond to simultaneous queries.
- Tag collision occurs when many tags are present in a small are

SPECIAL APPLICATIONS

General

Travel time, or the time required to traverse a route between any two points of interest, is a fundamental measure in transportation. Travel time is a simple concept understood and communicated by a wide variety of applications for transportation engineers and planners. Several data collection techniques can be used to collect travel times. These techniques are designed to collect travel times and average speeds on designated roadway segments or links.

Travel Time Data collection Technique

Following are the different techniques available for the travel time data collection.

- Test Vehicle Techniques
- License Plate Matching Techniques
- ITS Probe Vehicle Techniques
- Emerging and Non-Traditional Techniques

Test Vehicle Techniques

Travel time data using active test vehicles in combination with varying levels of instrumentation: manual (clipboard and stopwatch), an electronic distance measuring instrument (DMI), or a global positioning system (GPS) receiver.

It involves the use of data collection vehicle within which an observer records cumulative travel time at predefined checkpoints along a travel route. Then this information converted to travel time, speed, and delay for each segment along the survey route. There are several different methods for performing this type of data collection, depending upon the instrumentation used in the vehicle. These vehicles are instrumented and then sent into the field for travel time data collection, they are sometimes referred to as "active" test vehicles.

Advantages

- Advanced test vehicle techniques (*e.g.*, DMI or GPS use) result in detailed data.
- Low initial cost.

Disadvantages

- Sources of possible error from either human or electric sources that require adequate quality control,
- Data storage difficulties.

License Plate Matching Techniques

Travel times by matching vehicle license plates between consecutive checkpoints with varying levels of instrumentation: tape recorders, video cameras, portable computers, or automatic license plate character recognition.

Advantages

- Travel times from a large sample of motorists, very simple technique.
- Provides a continuum of travel times during the data collection period.

Disadvantages

- Travel time data limited to locations where observers or video cameras can be positioned;
- Limited geographic coverage on a single day
- Accuracy of license plate reading is an issue for manual and portable computer

ITS Probe Vehicle Techniques

Travel times using ITS components and passive probe vehicles in the traffic stream equipped with signpost-based transponders, automatic vehicle identification (AVI) transponders, ground-based radio navigation, cellular phones, or GPS receivers. Some vehicles are equipped with dynamic route guidance (DRG) device which act as roving traffic detectors, a non-infrastructure based traffic monitoring system. Such vehicles, which are participating in the traffic flow and capable of determining experienced traffic conditions and transmitting these to a traffic center, are called probe vehicles. To determine its position and to register experienced traffic conditions, a probe vehicle is equipped with on-board electronics, such as a location and a communication device. By means of the location device, the probe vehicle keeps track of its own geographic position. Through the communication device, the probe vehicle transmits its traffic experiences via a mobile communication link to a traffic center. For instance, each probe can transmit traffic messages once every time interval containing its location and its speed at the instant of transmission. In this traffic center the traffic data received from probe vehicles is gathered, and combined with data from the other monitoring sources, and processed into relevant traffic information. It is very useful for Advanced Traveller Information system (ATIS).

Advantages

- Low cost per unit of data
- Continuous data collection
- Automated data collection
- Data are in electronic format
- No disruption of traffic

Disadvantages

- High implementation cost
- Fixed infrastructure constraints - Coverage area, including locations of antenna
- Requires skilled software
- Not recommended for small scale data collection efforts.

Index